RETARGETABLE CODE GENERATION FOR DIGITAL SIGNAL PROCESSORS

RETARGETABLE CODE GENERATION FOR DIGITAL SIGNAL PROCESSORS

Rainer **LEUPERS**
University of Dortmund
Department of Computer Science
Dortmund, Germany

KLUWER ACADEMIC PUBLISHERS

BOSTON / DORDRECHT / LONDON

A C.I.P. Catalogue record for this book is available from the Library of Congress

ISBN 0-7923-9958-7

Published by Kluwer Academic Publishers,
P.O. Box 17, 3300 AA Dordrecht, The Netherlands.

Sold and distributed in the U.S.A. and Canada
by Kluwer Academic Publishers,
101 Philip Drive, Norwell, MA 02061, U.S.A.

In all other countries, sold and distributed
by Kluwer Academic Publishers Group,
P.O. Box 322, 3300 AH Dordrecht, The Netherlands.

Printed on acid-free paper

All Rights Reserved
© 1997 Kluwer Academic Publishers
No part of the material protected by this copyright notice may be reproduced or
utilized in any form or by any means, electronic or mechanical,
including photocopying, recording or by any information storage and
retrieval system, without written permission from the copyright owner.

Printed in the Netherlands

CONTENTS

FOREWORD vii

PREFACE ix

1 INTRODUCTION 1
  1.1 Design automation for VLSI 1
  1.2 HW/SW codesign of embedded systems 2
  1.3 Embedded software development 5
  1.4 DSP algorithms and architectures 8
  1.5 Problems and solution approach 14
  1.6 Overview of related work 18
  1.7 Goals and outline of the book 27

2 PROCESSOR MODELLING 29
  2.1 MIMOLA language elements 29
  2.2 The MSSQ compiler 33
  2.3 Application studies 36

3 INSTRUCTION-SET EXTRACTION 45
  3.1 Processor description styles 45
  3.2 Analysis of control signals 48
  3.3 Binary decision diagrams 51
  3.4 Instruction-set model 52
  3.5 Internal processor model 59
  3.6 Behavioral analysis 60
  3.7 Structural analysis 68
  3.8 Postprocessing 78

	3.9 Experimental results	80
	3.10 ISE as a validation procedure	82
4	**CODE GENERATION**	85
	4.1 Target architecture styles	85
	4.2 Program representations	86
	4.3 Related work	89
	4.4 The code generation procedure	91
	4.5 DFL language elements	95
	4.6 Intermediate representation	98
	4.7 Code selection by tree parsing	105
	4.8 RT scheduling	118
5	**INSTRUCTION-LEVEL PARALLELISM**	127
	5.1 Address generation in DSPs	127
	5.2 Generic AGU model	130
	5.3 Addressing scalar variables	132
	5.4 Arrays and delay lines	146
	5.5 Code compaction	161
6	**THE RECORD COMPILER**	179
	6.1 Retargetability	179
	6.2 Code quality	184
7	**CONCLUSIONS**	191
	7.1 Contributions of this book	191
	7.2 Future research	193
REFERENCES		195
INDEX		207

FOREWORD

According to market analysts, the market for consumer electronics will continue to grow at a rate higher than that of electronic systems in general. The consumer market can be characterized by rapidly growing complexities of applications and a rather short market window. As a result, more and more complex designs have to be completed in shrinking time frames.

A key concept for coping with such stringent requirements is re-use. Since the re-use of completely fixed large hardware blocks is limited to subproblems of system-level applications (for example MPEG-2), flexible, programmable processors are being used as building blocks for more and more designs. Processors provide a unique combination of features: they provide flexibility and re-use.

The processors used in consumer electronics are, however, in many cases different from those that are used for screen and keyboard-based equipment, such as PCs. For the consumer market in particular, efficiency of the product plays a dominating role. Hence, processor architectures for these applications are usually highly-optimized and tailored towards a certain application domain. Efficiency is more important than other characteristics, such as regularity or orthogonality. For example, multiply-accumulate instructions are very important for digital signal processing (DSP). Most DSP processors use a special register (the accumulator) for storing partial sums. Orthogonality would require any register to be usable for partial sums, but this would possibly require another register field in the instruction and potentially lengthen the critical path for that instruction. In the light of the importance of efficiency, compiler techniques should be extended to handle features that contribute to the efficiency of embedded processors.

The contribution by Rainer Leupers is one of the few recent contributions aiming at providing efficient (in the sense of efficient code) compilers for embedded processors. These contributions show that efficient compilation is feasible, if the special characteristics of certain application domains (such as DSP) are exploited.

In this contribution, Rainer Leupers describes, how code generators for compiling DSP programs onto DSP processors can be built. The work includes a number of optimization algorithms which aim at making high-level programming for DSP applications practical. It can be expected that these and similar optimizations will be included in many future compilers for DSP processors. It can also be expected that the whole area of optimizing techniques will receive an enormous attention in the future, possibly also stimulated by Leupers' contribution.

A key aspect of the approach developed by Rainer Leupers is the automatic generation of code generators from hardware descriptions of the target processor. Hence, retargetability is achieved. Retargetability is the ability of quickly generating new tools for new target processors. Why is retargetability desirable? Due to the interest in efficient processors, domain or even application-specific processors are of interest. This means that a significant number of different architectures and instructions sets is expected to exist. Retargetability makes the generation of compilers for these architectures economically feasible.

In the approach developed by Leupers, compilers can be generated from a description in a hardware description language. This approach is hence the first that really bridges the gap between ECAD and compiler worlds. This reduces the amount of component models that are required and enables new applications such as analyzing the effect of hardware changes with respect to different cost and performance metrics. This also enables a better understanding of the software/hardware interface.

According to common belief retargetable compilers are not as efficient as target-specific compilers. Nevertheless, Leupers' RECORD compiler outperforms a target-specific compiler for a standard DSP on the majority of the benchmark examples.

The design of the compiler generation system required background knowledge about similar previous work. Some of this knowledge was available in the department Leupers was affiliated with. Much of this knowledge came from earlier projects on compiler generators. Some key concepts such as the concepts of instruction conflicts and alternative code covers could be kept. A huge amount of others are new. It therefore makes me happy to know that these solutions will be generally available in the form of this book.

Dortmund, April 1997   P. Marwedel

PREFACE

This book responds to the demand for a new class of CAD tools for embedded VLSI systems, located at the edge between software compilation and hardware design. Compilation of efficient machine code for embedded processors cannot be done without a close view of the underlying hardware architecture. In particular this holds for digital signal processors (DSPs), which show highly specialized architectures and instruction sets. Very few attention has been paid so far to development of high-level language compilers for DSPs, which is also expressed in the following statement by Hennessy and Patterson [HePa90]:

> "Digital signal processors are not derived from the traditional model of computing, and tend to look like horizontal micropro-grammed machines or VLIW machines. They tend to solve real-time problems, essentially having an infinite-input data stream. There has been little emphasis on compiling from programming languages such as C, but that is starting to change. As DSPs bend to the demands of programming languages, it will be interesting to see how they differ from traditional microprocessors."

As a result, the code quality achieved by commercial compilers for DSPs is still far from satisfactory. In addition, current compiler technology does not support frequent changes of the target processor, which would be necessary for effective hardware-software codesign environments. In this book, new techniques for retargetable and optimizing compilers for DSPs are presented and put into context with related work. The whole compilation process is covered, including target processor capture, intermediate code generation, code selection, register allocation, scheduling and optimizations for parallelism.

This research monograph is a revised version of my doctoral thesis, which has been submitted to the Department of Computer Science at the University of Dortmund (Germany) in November 1996. I would like to thank my advisor Prof. Dr. Peter Marwedel and my co-referee Prof. Dr. Ernst-Erich Doberkat for their efforts and valuable comments. Furthermore, I would like to thank my colleagues at the "LS XII" division of the Computer Science department at the University of Dortmund. Particular support has been provided by Steven Bashford, Birger Landwehr, Wolfgang Schenk, and Ingolf Markhof. Dr. Detlef Sieling gave important hints concerning binary decision diagrams. I also gratefully acknowledge the help of Dr. Jef van Meerbergen (Philips Research Labs, Eindhoven, The Netherlands) and Vojin Zivojnovic (Technical University of Aachen, Germany), who contributed material for application studies.

Last but not least, I would like to thank my family for providing some technical and so much mental support and encouragement. I dedicate this book to my ladies Bettina, Helen, and Paulina.

Dortmund, April 1997   Rainer Leupers

1
INTRODUCTION

1.1 DESIGN AUTOMATION FOR VLSI

Advanced silicon technologies permit implementation of digital circuits comprising millions of transistors. In order to keep design of such complex systems manageable, the design process is commonly subdivided into several views and levels of abstraction, which can be visualized by the well-known Y-chart (fig. 1.1). Starting from an abstract, functional view of the system-under-design, the goal is construction of the *physical layout*, i.e. the set of documents needed for chip fabrication. Electronic computer-aided design (ECAD) tools aim at automating design steps represented by edges between crosspoints in the diagram. Traditionally, ECAD technology was mainly concerned with single hardware components. Practical electronic systems, however, are often composed of several components. These may not only include special-purpose hardware, but also user-programmable "off-the-shelf" processors. Problems arising at this high level of abstraction, e.g. partitioning a complete system into several chips and selecting appropriate processors, are not yet solved by commercial tools. Consequently, ECAD research is currently taking the step towards system-level design automation.

Economically reasonable VLSI systems in general consist of both hardware and software components. Parts of the system, which are not subject to tight computation-speed constraints, are better realized by software running on programmable processors, while other parts might require fast, dedicated hardware. It is advantageous to implement as much of a system as possible in software, because software is cheaper and more flexible. The trend towards VLSI systems comprising both hardware and software components creates a need for software compilers *as a part of ECAD systems*. Obviously, compilers

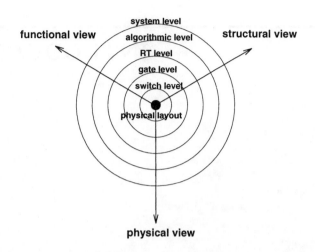

Figure 1.1 The Y-chart

that map high-level language programs into machine code for general-purpose processors are among the most widespread software tools and have reached a high degree of maturity. However, current compiler technology does not meet the requirements that arise in design of combined hardware-software systems. In general, this is due to the presence of special-purpose processors and very high code quality requirements. Moreover, software compilers in ECAD environments need to be linked to hardware design tools and must provide a certain flexibility. In particular, this holds for the design of *embedded systems*.

1.2 HW/SW CODESIGN OF EMBEDDED SYSTEMS

Electronic computer systems can be roughly divided into two classes: general-purpose and special-purpose systems. General-purpose systems like PCs may be programmed and configured by the user, so as to serve a variety of different computer applications, such as information systems or scientific computations. In contrast, special-purpose systems are designed and installed only once to serve a single, particular application. They interact with larger, often non-electronic environments and are therefore commonly called *embedded systems*. Today, embedded systems are found in many areas of everyday life, such as telecommunication and home appliances. Besides small physical volume and

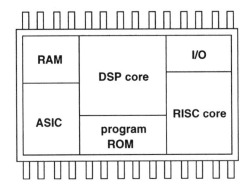

Figure 1.2 Heterogeneous single-chip embedded system

weight requirements, embedded systems usually must guarantee *real-time response*. In portable systems, also energy consumption may be of high importance.

Deep submicron VLSI technology permits to design complete systems on a single chip, resulting in higher speed and dependability at lower silicon area and power consumption [GCLD92, PLMS92]. VLSI systems comprising both HW and SW components are called *heterogeneous systems*. Fig. 1.2 shows a possible floorplan of a heterogeneous single-chip system. Software components of the system are executed on programmable *embedded processors*. Incorporation of processors as parts of a larger single-chip system is possible, since semiconductor vendors have made a number of existing packaged processors available in form of *cores*. Cores are layout macro-cells which can be instantiated by a design engineer from a component library. Further system components include memory modules, dedicated non-programmable hardware, and peripherals.

Starting from a purely functional description of a system-under-design, one of the first problems that need to be solved is *hardware-software partitioning*, i.e. the decision whether to assign the different components of the system to either hardware (ASICs) or software (cores). Fast ASICs need to be designed, whenever software cannot meet real-time constraints imposed on the system. On the other hand, special hardware should be minimized, in order to achieve low chip area. However, HW/SW partitioning is only one subproblem in embedded system design. A more general view is shown in fig. 1.3, commonly known as *hardware-software codesign*. Starting from a system-level specification, a partitioning step assigns parts of the system to either hardware or software. ASICs are generated for those parts which have been assigned to hardware,

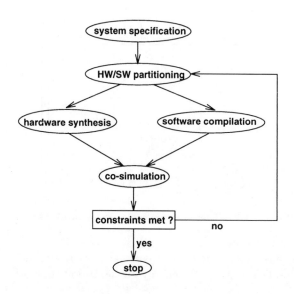

Figure 1.3 Hardware-software codesign flow

and the remaining parts are compiled into processor-specific machine code. Since the partitioning step is typically based on rough estimations, HW/SW co-simulation is necessary in order to verify that all constraints in the specification are met. In case of failure, re-partitioning is performed until a valid system implementation is found. Although the general procedure is simple, detailed realization of ECAD tools for HW/SW codesign faces difficult problems. Mainly, this concerns the areas of unified system specification languages, system architecture selection, accurate HW/SW performance estimation techniques, HW/SW co-simulation, and – the subject of this book – efficient code generation for embedded processors.

The state-of-the-art in HW/SW codesign for embedded systems is summarized in several books [Calv93, GVNG94, Gupt95, KAJW96]. A concise tutorial on HW/SW codesign is given by Gajski and Vahid [GaVa95].

1.3 EMBEDDED SOFTWARE DEVELOPMENT

The granularity of HW/SW partitioning is not necessarily restricted to the system level. Even after parts of the system have been assigned to software, a number of design decisions remain open with respect to the mapping of software to the instruction set of a programmable processor. This is due to the fact that embedded software development frequently is not a rigid process of compiling and debugging programs for a fixed target processor, but may also include the design of an *application-specific* instruction set. Such an extended view of embedded software development is illustrated in fig. 1.4, which comprises the following steps:

Instruction-set specification: If the target processor is not completely fixed in advance, it is necessary to develop a formal model which captures the behavior of the target machine and which can be used for the subsequent compilation and simulation steps. Initially, the designer has some intuition of a "suitable" instruction set, which can be derived from the application. For signal processing applications, for instance, this set will typically comprise instructions for fast execution of arithmetic operations.

Code generation: The source program is mapped to machine code in accordance with the specified instruction set. This can be done by assembly-level coding or – more efficiently – by high-level language (cross-)compilers. Obviously, in order to avoid the necessity of a separate compiler for each different instruction set, flexible *retargetable* compilers are necessary, which can be quickly adapted to new target processors. In addition, compilers for embedded software need to perform *optimizations* on the generated code, so as to meet the demands on very high code quality.

Profiling: Compiling the source program to a host computer and running it on a representative set of input data yields information on the typical execution frequency of program blocks. This permits to identify "hot spots" in the program, which provides hints for useful modifications of the source code. In combination with information on the compiled target machine code length for each block, profiling data can be used to obtain an estimation of whether the target machine code will meet the performance goal [LPJ97].

Simulation and debugging: In order to verify the correctness of the program specification and the code generation process, execution of the machine code needs to be simulated. For standard processors, simulators are usually

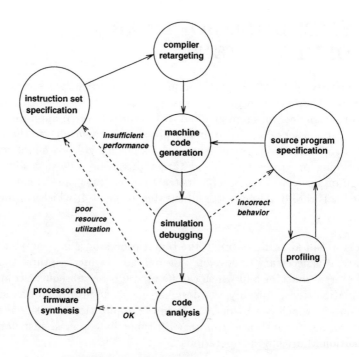

Figure 1.4 Embedded software design flow

available from the processor vendors. In case of application-specific instruction sets there is a need for retargetable simulators. Different methodologies have been published to implement such tools: In CBC [FaKn93] and CHECKERS [GLV96], a C language model for the target processor is automatically generated from the instruction-set model. This C model, which can be compiled onto a host machine, implements an interpreter for the target instruction set. A different approach is taken in the FlexWare system [PLMS95], making use of a generic retargetable VHDL model. The simulation speed achieved by these approaches is typically in the order of several kilo-instructions per second. A significant speed-up can be obtained by compiled simulation [ZTM95]. There, the target machine code is translated into an equivalent C program, which is compiled and executed on a host computer. Since this procedure eliminates the instruction decoding overhead of interpreter-based approaches, the simulation speed may reach more than a million instructions per second. The bottleneck, however, is the necessity to recompile the (possibly large) C simulation program after each change in the target machine code.

Also symbolic debugging of compiled machine code poses challenges on software development tools. Source-level debugging requires to retain the compiler symbol table information in the executable code. Particularly difficult is debugging of optimized machine code, in which case there is no longer a one-to-one correspondence between source statements and machine instructions. In special cases, this can be circumvented by attaching additional debug information to the machine code [ASU86]. However, when using the above-mentioned compiled simulation approach, symbolic debugging may be virtually impossible, because in this case it is necessary to maintain debug information across two hierarchy levels of compiler optimizations. A more pragmatic approach is to only use non-optimized code for the debugging phase. Nevertheless, debugging of optimized code cannot be fully avoided, since under worst-case conditions the behavior of optimized code may differ from its non-optimized counterpart.

Analysis of resource utilization: After verification of the compiled code, the specified instruction set can be synthesized in form of a programmable processor, and the machine code can be transformed into firmware for that processor. However, if the overall constraints on code size or speed are not met, only a deeper analysis provides hints for necessary modifications in the target instruction set. Even if the constraints are met, partial re-definition of the instruction set may lead to a cheaper implementation in terms of silicon area and/or power consumption. Analysis of resource utilization may be performed at two levels of abstraction: By analyzing the *utilization of instructions*, one can eliminate superfluous elements from the instruction set, so that the decoding logic and possibly also the instruction word-length may be reduced. Liem [LPJ97] has developed a tool for this purpose. A more fine-grained approach is to analyze the *utilization of components*, i.e. to keep track of the dynamic occupation of registers, functional units, and busses that implement the instruction set. This type of analysis enables HW/SW codesign at the processor architecture level. For instance, "moving" operations from software to hardware means duplicating a highly occupied functional unit in the processor data path. A code generation tool capable of component utilization analysis has been used at Philips [SMT+95].

From these considerations it should be obvious, that the compiler which generates target machine code plays a predominant role in this software development scenario. First of all, the compiler is responsible for the overall quality of the generated machine code in terms of code size and program execution speed. Furthermore, the results of resource utilization analyses are only meaningful, if the compiler is not the primary source of a possible waste of resources.

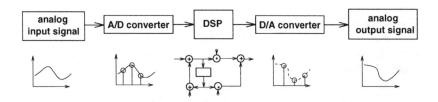

Figure 1.5 General digital signal processing system

In the remainder of this book we therefore concentrate on **code generation for embedded processors**, and **target processor modelling** (which includes instruction-set specification). In particular we consider **digital signal processors** (DSPs). DSPs are very common in embedded systems which involve a large amount of signal processing, such as in the areas of telecommunication and image processing. However, current compiler technology for DSPs is surprisingly poor. Traditionally, DSPs are regarded as processors *intended* to be programmed in assembly language, because of their highly specialized instruction sets and architectural features. While commercial compilers are available for some standard DSPs, the code quality is still far from satisfactory. Overheads of compiler-generated code of 500 % or more as compared to hand-crafted code have been reported [ZVSM94], which is mostly unacceptable. Typically, this overhead originates from the special characteristics of DSP applications.

1.4 DSP ALGORITHMS AND ARCHITECTURES

Both DSP algorithms and processor architectures show features which are hardly present in general-purpose computing. Here, we concentrate on DSP characteristics which are significant from a code generation viewpoint.

1.4.1 Digital signal processing

A general view of a digital signal processing system is shown in fig. 1.5. Input to the system is an analog signal, which might be generated by a microphone. After analog-to-digital conversion, the signal is processed, and the result is converted back to an analog signal, which in turn might drive a loudspeaker. Among the most important DSP applications is *digital filtering*. Digital filters

Introduction

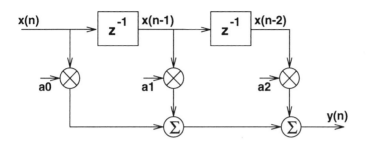

Figure 1.6 SFG of a finite impulse response (FIR) filter

modify a given input signal, so as to pass only certain parts of its frequency spectrum. They process discrete signals by performing arithmetic operations according to certain *filter algorithms*, commonly visualized by *block diagrams* or *signal flow graphs* (SFGs). The SFG of a *finite impulse response* (FIR) filter, is depicted in fig. 1.6. The filter operates in an infinite time loop. In each time step or *sample period* n the filter reads an input signal $x(n)$ and produces an output $y(n)$, which is computed by a set of multiplications and additions. The z^{-1} operator denotes a *unit delay*, i.e. a signal value from the previous sample period. The *filter coefficients* $a0, a1, a2$ characterize the filter transmission behavior. The functionality of the filter is given by the equation

$$y(n) = a0 \cdot x(n) + a1 \cdot x(n-1) + a2 \cdot x(n-2)$$

Although the FIR filter only represents a special case, it shows characteristics common to most DSP algorithms: Firstly, arithmetic operations are predominant in DSP, in particular sum-of-products or *multiply-accumulate* (MAC) computations, while control-flow operations, such as IF-constructs are rather rare. Secondly, DSP algorithms may refer to "delayed" signal values from previous sample periods. The set of delayed signals ($x(n-1)$ and $x(n-2)$ in the example) is called the *delay line* of an algorithm. Efficient realization of delay lines in software demands for an appropriate layout of data in memory.

An intuitive introduction to DSP theory and algorithms from an engineering perspective is to be found in [MaEw94]. DSP system design methodology including both hardware and software aspects is presented in [DLH88]. Lee and Messerschmitt [LeMe88] provide a comprehensive theory of digital communication.

1.4.2 DSP hardware architectures

The above-mentioned peculiarities of DSP algorithms as well as high computation speed requirements led to development of dedicated programmable DSP processors. Although showing significant differences in the detailed architecture, some characteristics are common to nearly all DSPs, distinguishing them from general-purpose processor types like CISCs and RISCs.

Common architectural characteristics

The primary goal in DSP design is *performance* with respect to arithmetic computations. In order to support fast execution of DSP algorithms, the following architectural characteristics are found in contemporary DSPs. Many of these have implications on code generation techniques.

Hardware multiplier: DSPs contain a dedicated hardware multiplier, capable of computing products within a *single machine cycle*. This eliminates the bottleneck of CISCs and RISCs, which often implement multiplication as a multi-cycle operation.

On-chip memory: In order to avoid wait states due to slow external memories, DSP chips typically comprise several kilobytes of on-chip RAM and ROM. On-chip RAM is often *distributed* over several memory banks accessible in parallel. Thus, appropriate assignment of data to different memory banks may increase utilization of the available memory bandwidth.

Address generation units: DSPs contain dedicated memory address generation units (AGUs) capable of performing address arithmetic independent of the central ALU. Efficient exploitation of AGUs during code generation significantly contributes to code quality.

Special-purpose registers: Typically, DSP data-paths are highly irregular, i.e. registers may be present anywhere in the data-path. The consequence of special-purpose registers on code generation is, that the instructions required to shuffle data from and to registers need to be explicitly taken into account.

Mode registers: Many DSPs employ a control mechanism in form of *mode registers*: Mode registers store control signals for combinational components, which need to be changed only rarely, thereby reducing the instruction word-length. For instance, mode registers are used for activation of certain arithmetic

Introduction

modes. Code generation must ensure that mode registers are loaded with appropriate values whenever a change in modes is required.

Ring buffers: Ring buffers simulate a *circular* address space in memory. They permit efficient implementation of delay lines in filters, but pose challenges on code generation: Variables in a DSP program which are candidates for ring buffers need to be identified, and circular addressing must be activated by special instructions.

Instruction-level parallelism: Presumably the most difficult task in DSP code generation is exploitation of potential parallelism. Most instructions of a DSP are executable within a single machine cycle, and usually a certain number (typically 3 to 6) of instructions can be executed in parallel. Identification of potential parallelism in a DSP program is mandatory for obtaining satisfactory code quality.

Extensive discussions of DSP characteristics and their historical development are to be found in [MaEw94], as well as in Lee's survey of programmable DSP architectures [Lee88].

Example

We now consider a real-life example: the Texas Instruments TMS320C25 fixed-point DSP [TI90]. Fig. 1.7 shows a simplified block diagram of its architecture. The processor data word-length is 16 bits, while ALU and accumulator internally operate on 32-bit numbers. There are on-chip memories for storing program and data. Memory addressing is done either directly or indirectly through an address register file. The AGU can execute address register updates in parallel to arithmetic operations. Address registers themselves are addressed by an address register pointer (ARP). Besides the accumulator register (ACCU) there are two special-purpose registers: TR stores operands for the hardware multiplier, and PR stores multiplier results, which can be fed to the ALU. Shifters are used for 16/32 bit number conversions, which are controlled by mode registers (not shown). The 16-bit instruction format of the TMS320C25 permits parallelization of several operations: As an example, consider the "MPYA" (*multiply and accumulate previous product*) instruction, having the following format in indirect addressing mode:

15	14	13	12	11	10	9	8	7	6	5	4	3	2 .. 0
0	0	1	1	1	0	1	0	1	IDV	INC	DEC	NAR	YYY

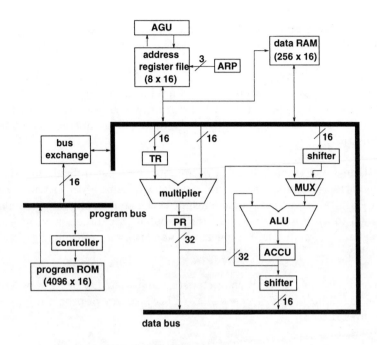

Figure 1.7 Simplified architecture of Texas Instruments TMS320C25 DSP

Bits 15 down to 7 constitute the fixed instruction *opcode*. MPYA executes two arithmetic operations simultaneously: Register TR is multiplied with a memory value accessed by an address register, and the result is stored in PR. In parallel, the previous contents of PR are added to the accumulator. Bits 6 down to 4 specify a possible update on the current address register, i.e. that register currently pointed to by ARP. The bits IDV, INC, and DEC determine whether or not the current address register is modified. An address register can be either incremented or decremented by 1, or by the contents of the distinguished address register number 0. Bit 3 (NAR) decides on an update of register ARP: If NAR = 1 then ARP is loaded with the concatenated values of bits 2 down to 0 (YYY).

This example gives an impression of the highly specialized instruction sets of DSPs. Obviously, such instructions are not easily generated from high-level language programs. Another major difficulty arises from the increasing use of *customizable* DSPs. This leads to a further classification of DSPs and the need for flexible compilers. Due to the limited scope of this book, we only

Introduction

focus on embedded *DSPs*. Marwedel's "cube model" [Marw95] gives a broader classification of embedded processors.

DSP classification

Many contemporary embedded systems involving DSP functionality (e.g. a cellular phone terminal) consist of a set of *packaged* processors, which are available as standard components from semiconductor vendors. Use of such **standard DSPs** leads to a comparatively low design effort. In terms of speed, power consumption and physical volume, however, separately packaged standard DSPs, having a *fixed architecture*, are not the most cost-effective solution. Therefore, future-generation systems will be realized as single-chip systems. While several standard DSPs (including the TMS320C25) are already available in form of cores, which can be used in single-chip systems, a particular application might not require the full amount of capabilities of a standard DSP. Using standard DSP cores thus leads to a possible waste of silicon area and energy consumption. As a consequence, system houses are starting to use *flexible, customizable DSPs*. For these processors, the term **ASIP (application-specific instruction-set processor)** has been created. In contrast to standard DSPs, ASIPs are usually *not packaged*. Frequently, ASIP architectures can be parameterized, e.g. by the size of register files or the number of functional units in the data-path. Only the coarse architecture is fixed in advance, so that a chip designer may trade silicon area against computation speed. The ASIP can be tailored towards a specific application by iteratively re-mapping of program sources onto different detailed architectures.

Standard DSPs are mostly available with software development tools. While compilers standard DSPs suffer from insufficient code quality, the situation for ASIPs is even worse: Since ASIPs are *in-house devices*, only used for a small number of applications before becoming obsolete, there is hardly any high-level language compiler support for ASIPs. Insufficient quality and non-availability of compilers lead to the fact that even nowadays DSPs are mostly programmed in assembly languages, which implies all the well-known disadvantages of low-level programming. As time-to-market is now the most important issue for VLSI system houses, taking the step towards high-level programming seems mandatory.

1.5 PROBLEMS AND SOLUTION APPROACH

The unsatisfactory situation in current software compiler technology for DSPs leads to consideration of two main problems in this book:

Improving retargetability: Compilers for embedded DSPs need to offer flexibility. They should be easily *retargetable* towards different processors or at least different parameterizations of a generic ASIP. Retargetability permits to study the mutual dependence between hardware structures in a processor and program execution speed, so as to find the most cost-effective system implementation under given area, performance, or power-consumption constraints. In a HW/SW codesign process as depicted in fig. 1.3, a retargetable compiler would be a valuable tool for trying out different options for target processors, to which the software components of an embedded system could be mapped. Existing HW/SW codesign systems, such as VULCAN [GuDe92], CHINOOK [ChBo94], COSYMA [EHB93], and CODES [BSV93], only target some off-the-shelf processors like R3000, SPARC, and 8086. They emit C code which is translated by processor-specific compilers. The PTOLEMY system [KaLe93] directly generates assembly code for Motorola DSPs. So far, system architecture exploration by means of retargetable compilers is not provided. The same holds for commercial systems, such as DSPStation (Mentor Graphics/EDC) and COSSAP (Synopsys), which only support a set of standard DSPs. A recent HW/SW partitioning project aims at closer cooperation with retargetable compilers [NiMa96].

Improving code quality: The poor code quality of current DSP compilers is essentially caused by the fact, that only classical compiler construction techniques have been adopted for DSPs. The highly dedicated instruction sets and instruction formats of DSPs, however, demand for code generation techniques beyond the scope of CISC or RISC compilers. Using high-level language compilers instead of assembly-level programming for DSPs will only gain industrial acceptance, if compiler-generated code does not incorporate a significant overhead compared to hand-crafted code.

Problems related to embedded code generation tools are also summarized in [Marw95, ADK+95]. Vanoostende et al. [VVE+95] consider the relevant problems from an industrial perspective.

1.5.1 Trade-offs in compiler construction

As with most real-world problems, the above goals are contradicting. The more a compiler is tailored towards a certain target processor, the higher is the code quality and vice versa. In fact, the trade-offs that can be made when designing a retargetable code generator for DSPs are represented by a four-dimensional search space: In addition to retargetability and code quality, the acceptable amount of *compilation time* must be taken into account. It is important to note, that compilation time in DSP code generation is not of the same importance as in general-purpose computing. While the user of a C compiler on a workstation is hardly willing to spend more than a few minutes in compiling thousands of source code lines, the benefits of high-quality embedded DSP code justify much higher compilation times. Even running a compiler on a DSP program "overnight" may be acceptable in some cases. Furthermore, DSP programs are typically rather short compared to usual software. The fourth dimension is the amount of *user interaction*: In contrast to processor-specific compilers in general-purpose systems, retargetable compilers cannot act as push-button tools, but for instance require assistance through externally specified translation rules, that partially guide code generation.

The way DSP code generation is approached in this book is sketched in fig. 1.8. The algorithm to be compiled is described in a high-level programming language. Retargetability is achieved by using an external target processor model specified by the user. Both program source code and processor model are passed through frontends to the code generator. The code generator maps the source program into a machine program for the specified target processor, making use of a user-defined transformation library. The following sections describe the input formats for software and hardware in more detail.

1.5.2 Programming language

General-purpose programming languages do not adequately support specification of DSP algorithms. While other approaches favor rather peculiar "extended subsets" of the C language [Inte96], we use a programming language designed for DSPs: the **Data Flow Language** (DFL). DFL [DFL93] is the successor of the **SILAGE** language. SILAGE has been developed at the University of California at Berkeley as a specification language for DSP algorithms [Hilf85]. Extended by imperative language elements, DFL is the commercial version of SILAGE. Mentor Graphics offers the DFL-based DSP synthesis system "DSPStation". DFL is a high-level programming language comprising

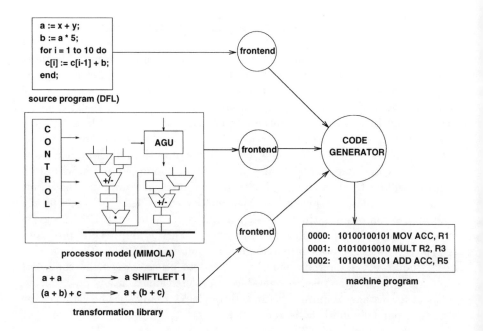

Figure 1.8 Functionality of the retargetable code generation system

control structures like conditional statements and loops. The salient features of DFL compared to general-purpose programming languages are:

- DFL captures the pure *data-flow* of DSP programs, and is therefore well-suited for description of signal flow graphs.
- DFL permits *bit-true* specifications, i.e. the exact bit-widths of data types can be defined by the user.
- DFL provides *DSP-specific operators*, which support modelling of delay lines and make characteristics of arithmetic operations explicit.

1.5.3 Target processors

As we will discuss in chapter 3, a versatile retargetable compiler should support both structural and behavioral processor models, and even mixtures between those. Furthermore, the necessary link to hardware design tools must be taken into account. It is therefore reasonable to use a *hardware description language*

Introduction

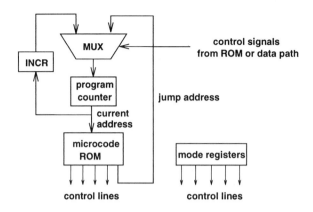

Figure 1.9 Microprogrammable controller architecture

(HDL) for processor modelling. HDLs cover both behavioral and structural descriptions, and provide a close link to other design tools, e.g. for synthesis and simulation. In this book we use the **MIMOLA** HDL[1] for modelling target processors. MIMOLA permits modelling of arbitrary, programmable and non-programmable hardware. Its expressiveness ranges from the algorithmic to the gate level in the Y-chart. For the purpose of retargetable code generation for DSPs we consider the class of processors satisfying the following architectural specification.

The processor *controller architecture* must match the block diagram shown in fig. 1.9. Control signals steering all processor components originate either from a *microcode ROM* or from an (optional) set of *mode registers*. A *program counter* register points to the current program address in ROM. A multiplexer steered by control lines from ROM or data-path signals (flags) selects the next program address in each machine cycle. The next address is either a jump address stored in ROM or the current address incremented by one. Furthermore, we assume a single-phase global clocking scheme: Each primitive processor operation (*microoperation* or *register transfer*) takes place within a single machine cycle, and all sequential processor components (registers and memories) are updated at the end of a machine cycle. Due to this assumption we do not consider *floating-point* processors, which involve multi-cycle operations, but only *fixed-point* DSPs. We neither consider *instruction pipelining* visible to

---

[1] We prefer MIMOLA over the standard hardware description language VHDL [IEEE88, LSU89] for sake of simplicity. MIMOLA is sufficient for processor modelling, it is easy to learn, and it avoids the description overhead sometimes encountered in VHDL models. The concepts, however, are language-independent, and an equivalent VHDL-based frontend could be easily developed after definition of a suitable VHDL subset.

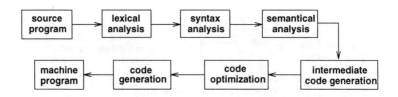

Figure 1.10 Structure of a "classical" compiler

the programmer. Retargetability within the scope of this controller model is guaranteed by permitting an *arbitrary, user-definable* instruction format.

Also the *data-path* of the target processor is completely user-definable. It consists of *combinational modules* (ALUs, multiplexers, decoders), *sequential modules* (registers, memories) and their interconnections in terms of wires and busses. The MIMOLA language permits modelling of a data-path either in terms of an explicit netlist, i.e. in a structural manner, or in an abstract behavioral fashion, hiding the internal structure. Selection of an appropriate modelling style is up to the user, who may choose the most suitable one for a given application. A complete target processor model in MIMOLA contains an integrated description of controller and data-path. An example is given in section 2.1.

1.6 OVERVIEW OF RELATED WORK

This section is intended to provide a coarse overview of relevant previous work and to draw conclusions on necessary improvements. We divide the overview of related work into four categories.

1.6.1 General compiler construction

The general structure of a compiler is shown in fig. 1.10. The source program, given in a high-level language, is first passed through **lexical and syntax analysis**. The theory of scanning and parsing context-free languages is well-developed, and practical results are available in form of tools like `lex` and `yacc`. Symbol tables in conjunction with attribute grammars also make **semantical analysis** a less complicated task.

Introduction

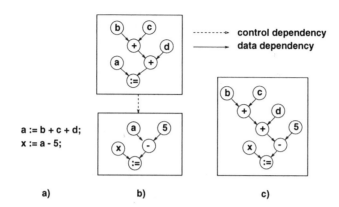

Figure 1.11 Intermediate representation: a) source code, b) graph-based representation (CDFG), c) CDFG after data-flow analysis

The most widespread **intermediate representations** (IRs) of programs are graph-based. The dependencies between program entities such as variables, operations, and assignments are made explicit in graph-based representations. Graph nodes correspond to program entities, and edges denote *control and data dependencies*. Such a representation is commonly called *control/data-flow graph* (CDFG). By means of *data-flow analysis*, source code statements can be composed to larger subgraphs, resulting in higher freedom for code optimization (fig. 1.11).

In the following **code optimization** step, machine-independent standard optimizations like *constant folding, strength reduction,* and *dead code elimination* are applied to the IR. Since many standard optimizations are well-known, these are are not further addressed in this book.

The most critical step in compilation is **code generation**, in which the IR is mapped to real machine instructions, while performing further optimizations. Code generation is often decomposed into three (partially overlapping) phases:

Code selection deals with selecting those instructions (from the set of all instructions available on the target machine), which are used to implement the desired program behavior. The goal is selection of a minimum number of instructions.

Register allocation decides, which program values will reside in registers at a certain point of time. Since registers and register files are typically critical resources in terms of storage capacity, an important subtask of register allocation is *register spilling*, i.e. temporarily releasing registers by storing their contents to (non-critical) memory. In general, register allocation deals with minimization of data move operations from and to registers.

Scheduling orders selected machine instructions in time while obeying possible inter-instruction dependencies. Scheduling ranges from *partial ordering* of instructions to detailed assignment of instructions to *control steps*. The primary goal of scheduling depends on the target processor family: In case of instruction pipelines visible to the compiler (mainly in RISCs) scheduling aims at minimization of pipeline conflicts. For processors with instruction-level parallelism (including DSPs), exploitation of potential parallelism is of major concern.

The main difficulties in code generation are twofold: First of all, even under strict separation of code generation phases, *optimal* code generation is NP-hard[2] in general. This has been shown for code selection, scheduling [GaJo79], and register allocation [Seth75] and a number of other compiler-related optimization problems. As a consequence, a vast amount of heuristics are currently used in code generation, and suboptimal solutions must be accepted. Furthermore, the above phases are mutually dependent, and each phase may profit from information generated by another phase. This results in the problem of *phase ordering*. However, no particular ordering of phases is clearly superior: Any permutation may outperform the others for certain processors and certain programs. Therefore, recent research on compiler construction aims at *phase coupling* in order to combat code quality overheads due to separation of phases.

The main reasons why general-purpose code generation techniques do not directly apply to DSPs are the following:

- The freedom for code selection is small in general-purpose processors: there are dedicated functional units, e.g. for integer or floating point operations. Frequently, DSPs comprise several identical functional units. Therefore,

---

[2] The class NP denotes the set of all decision problems that can be solved in polynomial time by a nondeterministic Turing machine. A decision problem Π is *NP-complete*, if $\Pi \in NP$, and Π is "at least as difficult" as all problems in NP, which can be formally proven by constructing a *polynomial-time transformation* between problems. In code generation, one usually has to deal with optimization problems. Optimization problems with an NP-complete decision counterpart are called *NP-hard*. For the detailed theory cf. [GaJo79]. In practice, both "NP-completeness" and "NP-hardness" imply that a problem is most likely to be intractable.

Introduction 21

code selection for DSPs must perform careful *resource binding* of operations. Moreover, operator chaining, for instance in form of MACs, must be taken into account.

- Techniques for register allocation assume a *homogeneous* register set, i.e. registers are interchangeable, so that mainly the *size* of register files is important. Special-purpose registers in DSPs are, however, not interchangeable, and register allocation cannot be separated from code selection.

- Scheduling techniques are still few and are primarily concerned with RISCs and superscalar processors, which essentially demand for re-ordering of instructions. Instruction-level parallelism has hardly been an issue.

- The machine models in compiler construction are very abstract. Capturing the specific architectural details of DSPs, however, demands for more *hardware-oriented* machine models. Code generation for DSP-specific hardware architectures has hardly been treated in the compiler community.

Several books provide overviews of compiler technology: [AhUl72] focus on the theory of formal languages and parsing. The standard reference for practical compiler construction is [ASU86] covering the whole compilation process except scheduling. Compilation techniques for different classes of programming languages are presented in [WiMa95], which also briefly touches scheduling problems. Discussions of techniques for single code generation phases are also given in doctoral theses [Brig92, Emme93]. Evaluation of existing compiler technology with respect to *irregular architectures* and *retargetability* has been the purpose of an exhaustive literature survey [Bash95].

1.6.2 Retargetable code generation

Within the scope of this book, it is sufficient to consider compilation of a *single* programming language to different target processors. Therefore, we use the following informal definition of retargetability:

1.6.2.1 Definition
A compiler for a fixed programming language is **retargetable**, if it can be adapted, so as to generate machine code for any processor within a defined class of processors, in such a way that the largest part of the compiler source code is retained.

It is also important to distinguish different *levels* of retargetability:

Processor-specific: A fixed target processor model is hard-coded in the compiler, and the code generation techniques only work for that target processor. Most of the compiler source code would need to be rewritten for another processor.

Portable: The code generation methods in the compiler are general enough, so that the compiler can be adapted to significantly different target processors by its developers, typically by rewriting compiler source code.

User-retargetable: The compiler uses an *external* target machine description in a compiler-specific description language, which can be edited by the user. Retargeting the compiler may imply rewriting parts of its source code, so as to provide processor-specific hints to the compiler.

Machine-independent: The compiler uses an *external* target machine description, which can be edited by the user. All instruction-set details needed for code generation are automatically derived from the machine description. Rewriting compiler source code is not necessary.

Parameterizable: The compiler works only for a narrow class of processors having a common overall architecture. An external machine description is used, which only consists of *numerical parameters* such as word-lengths, register file sizes and the number of functional units. Rewriting compiler source code is not necessary.

For the intended area of "HW/SW codesign of embedded systems", which demands for short turnaround times, only the latter three are of interest.

One of the earliest contributions regarding user-retargetable compilers goes back to Glanville [Glan77]. He proposed to use LR (shift-reduce) parsing of source code statements with respect to an instruction-set grammar. Satisfactory results were reported for some machines, but instruction-level parallelism was not treated. Although relying on a well-defined formal background, the LR parser based approach suffers from the ambiguity of grammars and therefore leads to suboptimal code selection on parallel machines. Sometimes, ambiguous grammars can be transformed into non-ambiguous ones, but this is apparently beyond the scope of user-retargetability. Cattell [Catt78] proposed a heuristic target-independent code selection method. However, his machine description formalism neither captured parallelism. The survey by Ganapathi et al. [GFH82] summarizes the techniques available in the early eighties. The

Introduction

GNU C compiler gcc [Stal93] could be successfully retargeted to a number of CISC and RISC machines. Unfortunately, gcc is located at the edge between portable and user-retargetable compilers, since it requires an exhaustive target machine description in a very specific language.

Different approaches to retargetable *peephole optimization* and scheduling have been proposed [DaFr84, Alla90, BHE91]. Peephole optimization denotes rule-based, local code optimization after code generation. Partial retargetability with respect to code selection is provided by *code generator generators*: Tools like BEG [ESL89], Twig [AGT89], and iburg [FHP92a] are capable of generating fast processor-specific tree parsers from instruction-set descriptions given as *tree grammars*.

A special area is retargetable generation of *self-test programs*. The MSST [Krug91] and RESTART [BiMa95] systems generate binary test programs based on structural processor models. Due to their limited scope, however, both systems pay less effort to code optimization.

1.6.3 Microprogramming

Among the different processor families, microprogrammable processors (MPPs) represent the lowest level of user-programmability: A loadable microcode memory gives explicit access to control lines of processor components. Therefore, only the programmer (or the compiler) is responsible for exploitation of parallelism. The instruction format of an MPP is subdivided into several *instruction fields* (fig. 1.12). Each field occupies some of a total of n instruction bits and controls exactly one data-path component. This is commonly called a *horizontal* instruction format. In other variants, less flexibility is provided due to *instruction encoding* aiming at instruction word-length minimization. Some MPPs also make use of *residual control*. In this case, more "static" control information is shifted to dedicated registers rather than being part of the instruction word. Residual control corresponds to mode registers in DSPs. Another characteristic of MPPs is the fixed *instruction length*, namely a length of one word. The following terminology is important with respect to MPPs:

1.6.3.1 Definition
A **microoperation** or **register transfer** (RT) is a primitive, atomic operation on an MPP. An RTs reads inputs from a set of sequential components or processor ports, performs a computation on these inputs, and stores the result

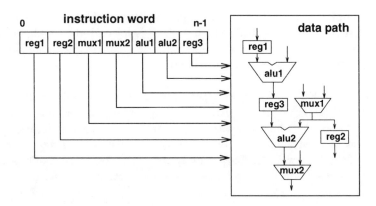

Figure 1.12 Instruction format in microprogrammable processors

into a sequential component within a single machine cycle. A **microinstruction** is a set of RTs, that are executable in parallel on an MPP. An MPP shows **instruction-level parallelism** (ILP), if its instruction set comprises at least one microinstruction that consists of more than one RT.

A number of techniques for exploitation of potential parallelism on horizontal (VLIW-like) machines have been developed. The general idea is to first translate source code into *vertical machine code*, consisting of separate, partially interdependent RTs. Afterwards, RTs are rearranged to form valid microinstructions in a *(micro)code compaction* phase, which can be regarded as low-level scheduling. Compaction algorithms fall into two classes: *Local* compaction techniques only handle *basic blocks*, i.e. vertical code segments having at most one jump or branch operation at their end. In contrast, *global* compaction also optimizes beyond block boundaries. Being a resource-constrained scheduling problem, already optimal local compaction is NP-hard [DeWi76]. A number of heuristic local compaction techniques turned out to be useful in practice [DLSM81]. A popular global compaction technique is Trace Scheduling [Fish81] which, unfortunately, tends to significantly increase the *code size* due to "compensation code" that needs to be inserted. Percolation scheduling [AiNi88] is another well-known technique for global compaction.

Different researchers have treated user-retargetable generation of microcode from high-level programming languages. In the MPG system [BaHa81], focus was on microcode generation for mainframes. Vegdahl [Vegd82a, Vegd82b] emphasized the necessity of phase coupling in microcode generation. Mueller

Introduction 25

and Varghese [MuVa83] proposed code generation based on a *structural model* of the target machine. However, much user interaction was required. More recent techniques, such as *loop folding* and *loop unrolling* aim at increasing potential parallelism through loop restructuring [Lam88, JoAl90].

1.6.4 Embedded code generation

A number of approaches to code generation for embedded DSPs have been published recently. Without considering detailed code generation techniques here, we list the significant contributions in chronological order. More detailed discussions are given where appropriate in the remaining chapters.

1987: Nowak's machine-independent MSSQ compiler [Nowa87a] generates microcode based on structural processor descriptions in the MIMOLA language. A graph model is used for internal representation of the target processor. Results are reported for different real-life processors, including ASIPs and standard DSPs. MSSQ suffers from insufficient code quality, but is a highly versatile tool in terms of retargetability. Since the contributions of this book are based on experiences with MSSQ, a separate description is given in section 2.2.

1988: Rimey and Hilfinger [RiHi88] introduce the concept of *data routing* in code generation for ASIPs. After operations have been bound to functional units in the target processor, data routing deals with transporting data between functional units, so as to minimize the amount of transport operations. Greedy scheduling orders operations in time based on information about "routability" of data. The target processor is described in a user-retargetable fashion by specification of available microoperations. Practical application is, however, restricted to a very simple class of target processors. A similar data routing technique was later presented by Hartmann [Hart92].

1992: Wess [Wess92] proposes DSP code generation based on *trellis diagrams*. His method integrates allocation of special-purpose registers into a linear time algorithm for code selection for arithmetic expressions. High-quality code generation is reported for standard DSPs. However, no mechanism is provided for constructing trellis diagrams from more common models. With respect to retargetability, the approach can thus be classified as being portable.

1993: Fauth's CBC compiler [FaKn93] is also based on data routing techniques. The nML language permits concise, hierarchical processor descriptions in a behavioral style. CBC uses a standard code generator generator for instruction selection and falls into the machine-independent category. CBC has been applied to a realistic design at Siemens. Marwedel [Marw93] reports details on MSSV, a former version of the MSSQ compiler. He also stresses the importance of retargetable compilers in the new area of HW/SW codesign.

1994: The CodeSyn compiler by Paulin et al. [LMP94a, LMP94b] maps C programs into machine code for industrial in-house ASIPs. Target processors are described by three separate items, identifying CodeSyn as a portable compiler: the set of available instruction patterns, a graph model representing the data-path, and a resource classification that accounts for special-purpose registers. Results are reported for some in-house ASIPs, for which high code quality was achieved. The CHESS compiler group at IMEC applies techniques of the former CATHEDRAL synthesis system to code generation for load-store architectures [LCGD94]. Wilson et al. study applicability of Integer Programming (IP) to code generation [WGHB94]. Their IP model permits complete phase coupling, and therefore generation of optimal code for basic blocks. Although the approach can be classified as user-retargetable, manual generation of Integer Programs requires much time, and practical application is only reported for a very restricted class of ASIPs. The "1st International Workshop on Code Generation for Embedded Processors" takes place at Schloß Dagstuhl, Germany. The resulting book [MaGo95] contains contributions from most research groups mentioned in this section.

1995: A research group at Philips describe a code generation methodology for industrial in-house DSP cores based on existing high-level synthesis tools [SMT+95]. A coarse ASIP architecture is first customized towards the given application by HLS supported by user interaction. This includes code selection and register allocation. A graph-based scheduling technique [TSMJ95] takes into account encoding restrictions and also time constraints. By taking advantage of the restricted target processor family, nearly optimal code is generated for a number of DSP algorithms. Research within the SPAM project focusses on code optimization rather than retargetability: Araujo and Malik partially integrate register allocation into tree parsing [ArMa95]. Due to usage of code generator generators, the approach is user-retargetable. Optimal code generation for data-flow trees is reported for the TMS320C25 DSP, however neglecting instruction-level parallelism and memory addressing. Liao and Wang et al. propose address assignment for DSP-specific AGUs as a means of advanced code optimization [LDK+95a], based on previous work by Bartley [Bart92]. They also investigate code selection on *directed acyclic graph* (DAG) intermediate representations [LDK+95c] and optimization of mode register usage [LDK+95b]. High code quality is achieved, but the techniques are hardly

Introduction

retargetable and suffer from too high computational complexity. Liao's doctoral thesis [Liao96] provides more details on SPAM techniques.

1.7 GOALS AND OUTLINE OF THE BOOK

The work presented in this book is driven by the demand for flexible DSP code generators in HW/SW codesign for embedded systems. While most recent work on embedded code generation primarily concentrates on *code quality*, and therefore tends to be rather processor-specific, we also treat *retargetability* as a central issue, essentially aiming at a reasonable compromise between both. We present novel approaches and improvements of previous work, while also exploiting research results from other areas, in particular logic synthesis, compiler construction, and operations research. The presented techniques have been implemented in the form of the prototype compiler system RECORD (Retargetable compiler for DSPs). The organization of the book is as follows:

The processor modelling formalism in RECORD is the MIMOLA hardware description language. In **chapter 2** we give an introduction to MIMOLA, and we present application studies carried out with an earlier MIMOLA-based code generator, the MSSQ compiler. The experiences gained through these application studies provided the motivation for the new techniques that have been developed for RECORD.

Previous retargetable compilers operate on either behavioral or structural processor models. In order to overcome this restriction, **chapter 3** presents a novel and fast model analysis technique, which permits the use of behavioral, structural, and even mixed processor models. The extracted internal processor representation is independent of the modelling style and creates the basis for code generation.

Chapter 4 focusses on the tasks of code selection, register allocation, and scheduling. In order to cope with the irregular hardware architectures of DSPs, we integrate a code generator generator, which automatically constructs an optimal tree parser from the internal target processor representation. The generated tree parser is used to map an intermediate representation of DFL programs to processor-specific instructions. We show, how the tree parsing approach can be embedded into the overall retargetable code generation procedure.

Exploitation of potential parallelism in generated machine programs is the subject of **chapter 5**. In the first part we present new algorithms for *address generation and assignment* which aim at efficient utilization of DSP address generation units. These techniques are intended to increase potential parallelism in DSP machine programs. In the second part we introduce a novel method for *code compaction* based on Integer Programming. The compaction method handles a broad range of instruction formats and is capable of producing locally optimal results.

The capabilities of the RECORD compiler in terms of retargetability and code quality are experimentally evaluated in **chapter 6**. **Chapter 7** concludes with a summary and hints for further research.

2
PROCESSOR MODELLING

In this chapter we first give an overview of the MIMOLA hardware description language. Then, we briefly present the MSSQ tool, an earlier MIMOLA-based compiler operating on RTL structural models. Based on two application studies with real-life processors, we analyze strengths and weaknesses of the techniques used in MSSQ. The results of the application studies provide the motivation for the advanced processor modelling and code generation techniques in the remaining chapters.

2.1 MIMOLA LANGUAGE ELEMENTS

Originally, MIMOLA [BBH+94] is a combined hardware description and programming language. A MIMOLA *program* is given in form of an imperative language algorithm. The algorithm is described in a hardware-oriented superset of the PASCAL language. However, as argued before, imperative languages are not most appropriate choice for describing DSP algorithms, but data-flow languages are more favorable. Therefore, we use MIMOLA only for describing the *target processor hardware*.

MIMOLA hardware descriptions are *netlists*, which consist of *modules* and their interconnections. Interconnections between modules are either unidirectional *wires* or bidirectional *busses*. Modules are described by an *interface* and a *body*. The interface specifies the name of a module and its *I/O ports*. The body decides on the general type of the module: In case of *structural modules*, the body in turn is a netlist. In this way, concise *hierarchical* descriptions are possible. *Behavioral modules* are the leaf nodes of a structure. For these, the body describes the pure functionality. Behavioral modules may be combinational or sequential. The following example shows a combinational module in MIMOLA:

```
MODULE ALU (IN i1, i2: (15:0); OUT outp: (15:0); IN ctr: (1:0));
BEHAVIOR IS
 CONBEGIN
   outp <- CASE ctr OF
             0: i1 + i2;
             1: i1 - i2;
             2: i1 AND i2;
             3: i1;
           END;
 CONEND;
```

The interface specifies names and types of I/O ports. The latter are denoted by *bit vector types* of the form (<upper index>:<lower index>). Bit vectors are strings over $\{0, 1, x\}$, where x denotes an *undefined* value or a *don't care*. Furthermore, the *port modes* (**IN** or **OUT**) are declared. The body is constituted by a **CONBEGIN** ...**CONEND** construct, denoting concurrent execution of the enclosed *statements*. In the example, only one *signal assignment* is present: The **CASE** expression evaluates to a certain arithmetic expression on the input ports, dependent on the value of port **ctr**. The result is assigned to port **outp**. All expressions in assignments are based on a comprehensive set of built-in primitive operators. These include arithmetic (+, -, *, ...), logic (**AND**, **OR**, **NOT**, ...), and special (**SHIFT**, **CONCAT**, ...) operators. The semantics of these operators are defined in [BBH+94]. The following example shows a simple sequential module, namely a register:

```
MODULE Reg16bit (IN inp:(15:0); OUT outp:(15:0); IN enable:Bit);
BEHAVIOR IS
 VAR S: (15:0);
 CONBEGIN
   IF enable THEN S := inp;
   outp <- S;
 CONEND;
```

The behavior of module **Reg16bit** is specified by two concurrent statements: Firstly, the **IF** statement decides on the next state of *module variable* **S**. If the value of port **enable** is 0, the state of S is retained. For **enable** = 1, **S** is loaded with the value of port **inp**. Such variable assignments are supposed to

Processor modelling

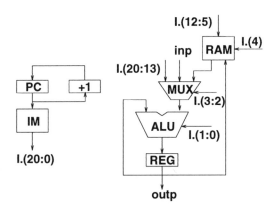

Figure 2.1 Schematic of a simple example processor. "I" denotes the instruction word.

be *synchronized by a global clock*. We assume that the maximum combinational delay in the processor is shorter than the clock period, so that variable states are updated with valid and stable new values at the end of each clock cycle. The second statement "outp <- S" describes assignment of the current value of S to port outp, independent of a control signal. By means of module variables, we can also more clearly specify the meaning of *register-transfer level*.

2.1.1 Definition
A **register-transfer level** (RTL) module is a behavioral module comprising at most one module variable.

Module interconnections are simply specified by *sources* and *sinks*, which are module ports. The corresponding syntactical construct is the CONNECTIONS statement, e.g.

```
CONNECTIONS    ALU.outp        ->   Reg16bit.inp;
               Reg16bit.outp   ->   ALU.i1;
```

Additionally, *bidirectional* connections can be declared by BUS statements. As an example for processor modelling, consider the schematics of a very simple processor in fig. 2.1, whose complete MIMOLA description is given in fig. 2.2.

```
MODULE SimpleProcessor (IN inp:(7:0); OUT outp:(7:0));
STRUCTURE IS                   -- outermost module is a structural one
TYPE InstrFormat = FIELDS      -- 21-bit horizontal instruction word
                 imm:      (20:13);
                 RAMadr:   (12:5);
                 RAMctr:   (4);
                 mux:      (3:2);
                 alu:      (1:0);
              END;
    Byte = (7:0); Bit = (0);   -- scalar types
PARTS                          -- instantiate behavioral modules
 IM: MODULE InstrROM (IN adr: Byte; OUT ins: InstrFormat);
     BEHAVIOR IS
     VAR storage: ARRAY[0..255] OF InstrFormat;
     CONBEGIN ins <- storage[adr]; CONEND;
 PC, REG: MODULE Reg8bit (IN data: Byte; OUT outp: Byte);
          BEHAVIOR IS
            VAR R: Byte;
            CONBEGIN R := data; outp <- R; CONEND;
 PCIncr: MODULE IncrementByte (IN data: Byte; OUT inc: Byte);
         BEHAVIOR IS
         CONBEGIN inc <- INCR data; CONEND;
 RAM: MODULE Memory (IN data, adr: Byte; OUT outp: Byte; IN c: Bit);
      BEHAVIOR IS
      VAR storage: ARRAY[0..255] OF Byte;
      CONBEGIN
        IF c then storage[adr] := data;
        outp <- storage[adr];
      CONEND;
 ALU: MODULE AddSub (IN d0, d1: Byte; OUT outp: Byte; IN c: (1:0));
      BEHAVIOR IS
      CONBEGIN               -- "%" denotes binary numbers
        outp <- CASE c OF %00: d0 + d1; %01: d0 - d1; %1x: d1; END;
      CONEND;
 MUX: MODULE Mux3x8 (IN d0,d1,d2: Byte; OUT outp: Byte; IN c: (1:0));
      BEHAVIOR IS
        CONBEGIN outp <- CASE c OF 0: d0;  1: d1; ELSE d2; END; CONEND;

CONNECTIONS
 -- controller:                   -- data path:
 PC.outp         -> IM.adr;       IM.ins.imm  -> MUX.d0;
 PC.outp         -> PCIncr.data;  inp         -> MUX.d1;    -- primary input
 PCIncr.inc      -> PC.data;      RAM.outp    -> MUX.d2;
 IM.ins.RAMadr   -> RAM.adr;      MUX.outp    -> ALU.d1;
 IM.ins.RAMctr   -> RAM.c;        ALU.outp    -> REG.data;
 IM.ins.alu      -> ALU.c;        REG.outp    -> ALU.d0;
 IM.ins.mux      -> MUX.c;        REG.outp    -> outp;      -- primary output
END; -- STRUCTURE
```

Figure 2.2   RT-level MIMOLA processor model

Processor modelling 33

Code generators operating on HDL models can hardly identify special components, such as the *program counter* (PC) and the *instruction memory* (IM) only by inspecting the structure. Therefore, MIMOLA permits to label PC and IM by means of *reservation statements*[1], which make the locations of PC and IM known to the code generator:

```
FOR PROGRAMCOUNTER USE PCRegister;
FOR INSTRUCTIONS USE InstrMemory;
```

2.2 THE MSSQ COMPILER

The MSSQ compiler is one component of the MIMOLA design system, a high-level ECAD environment for the design of digital programmable processors [MaSc93]. MSSQ compiles MIMOLA programs into microcode for predefined processors having a microprogrammable controller as defined in section 1.5.3. In contrast to other approaches, MSSQ derives all information required for code generation from an RTL netlist. In the following, we give an overview of the main data structures and code generation techniques in MSSQ. More detailed descriptions are provided in [Nowa87a, NoMa89, LeMa97].

2.2.1 Central data structures

The basic concept of code generation in MSSQ is quite simple: Pattern matching is performed between source code statements and processor structures, and the necessary control codes are collected "on-the-fly". The former is based on a graph data structure (the *connection operation graph, COG*), and the latter makes use of *instruction trees* (*I-trees*).

The COG is MSSQ's internal representation of the target processor, and it is constructed in a preprocessing step from the MIMOLA model. The COG *nodes* represent *hardware operators* and *module ports*, while *directed edges* represent *data-flow between nodes*. For variables in sequential modules, separate *read* and *write* operation nodes are present in the COG.

MSSQ assumes that all module control signals (directly or indirectly via decoders) originate at the labelled instruction memory. Therefore, control signals

---
[1] In case that a module has a unique module variable, we occasionally denote this variable simply by the module name.

are represented by *partial instructions*. For a target processor with instruction word-length W, a **partial instruction** or **version** is a bitstring $B \in \{0, 1, x\}^W$. I-trees are a data structure for *sets of versions*. Three operations are implemented on I-trees: SET constructs a single-node I-tree, and MERGE and CUT compute the union and the intersection of two version sets, respectively.

2.2.2 Preprocessing

MSSQ first transforms the source program into an RTL program. All user variables are bound to storage modules, and variable references are substituted by references to the corresponding storages. Furthermore, all high-level control structures like FOR, WHILE, and REPEAT loops are replaced by IF-constructs with explicit reference to the program counter register. IF-statements or IF-expressions are the only remaining high-level control structures. In hardware, IF-constructs correspond to multiplexers. Next, the COG is constructed from the hardware description. COG nodes representing hardware operations are annotated with the corresponding settings of module control ports, which can be derived from the module behavioral specifications.

2.2.3 Code generation

After source code transformation, the program to be compiled consists of a list of RTL assignments, each of which can be represented by a tree, whose root is the destination of the assignment. For each tree, MSSQ generates code in two sequential phases: The *allocation* phase comprises code selection, register allocation, and partially scheduling. Assignment of generated RTs to control steps is done in a heuristic *compaction* phase. The code sequences generated for each RTL assignment are finally concatenated, so as to form the complete machine program. An overview of this process is shown in fig. 2.3.

Allocation of an RTL assignment consists of *pattern matching* and *temporary allocation*. During pattern matching, MSSQ looks for a subgraph (an RT) in the COG, which matches the tree representation of the assignment. Then, MSSQ collects all module control signals required for that RT by means of I-trees. If pattern matching fails, temporary allocation is initiated by splitting the RTL assignment into a sequence of two simpler ones. A temporary location is selected, and pattern matching is recursively called for both new RTL assignments.

Processor modelling

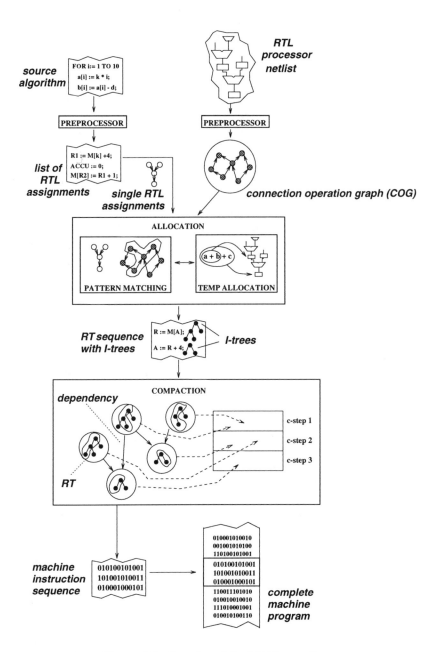

Figure 2.3 Overview of the MSSQ compiler

The compaction phase aims at exploiting potential instruction-level parallelism, thereby completing the scheduling phase. Compaction operates on the output from allocation, i.e. a sequence of RTs, each with an I-tree representing all its alternative versions. MSSQ employs a modified version of the "first-come first-served" (FCFS) heuristic [DLSM81]: The microinstructions are generated step-by-step, starting with the "last" control step. When "finishing" a microinstruction, MSSQ inserts additional partial instructions in order to avoid *undesired side effects*. Two sources are potentially responsible for such side effects: Whenever registers contain live values, but are not referenced in a certain microinstruction, then it must be ensured that its value is retained during that control step. MSSQ implements this by scheduling "no-operations" (NOPs). Similar to regular hardware operations, NOPs are associated with partial instructions. Furthermore, all modules driving a bidirectional bus must provide a "tristate" operation. In the simplest case, such modules are *tristate drivers*. These modules ensure, that bus conflicts can be avoided by disconnecting the driver from the bus. Similar to NOPs, MSSQ packs tristate operations for all unused bus drivers into each microinstruction.

The final output of MSSQ is the binary machine program listing, which corresponds to the required contents of the instruction memory. In the next two sections, we present results of two application studies carried out with MSSQ. In the first one focus is on the trade-off between code quality and user interaction. In the second one we concentrate on the task of processor modelling.

2.3 APPLICATION STUDIES

2.3.1 A digital audio ASIP

The first application study stems from the area of digital audio signal processing. The source algorithm is given by approximately 100 lines of SILAGE code, and describes a *digital bass booster* (DBB). The DBB consists of two identical stereo channels, each realizing a *low-pass filter*. In each sample period, the filter reads an input signal **in** and produces an output **out** by computing a series of multiplications and additions. Three delay lines are present: one of length six and two of length two.

The target processor for this DSP algorithm is a predefined in-house ASIP designed for digital filtering. Its schematic is depicted in fig. 2.4. The ASIP has a 41-bit microinstruction format, a dedicated address generation unit supporting

Processor modelling

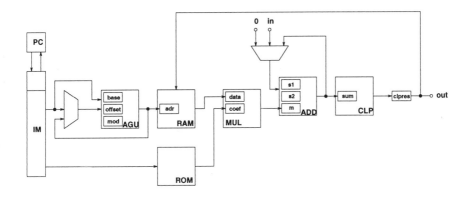

Figure 2.4 Schematics of DBB target processor

ring buffers, and a 120 × 18 bit RAM. Arithmetic computations are performed on a 24-bit multiply-accumulate section. Filter coefficients are stored in a ROM. All functional units can work in parallel in order to guarantee high throughput.

Processor modelling

The instruction format of the ASIP is specified as follows:

40:32	31:30	29	28:26	25	24:22	21:18	17:11	10:0
next PC	RAM	MUL	ADD	CLP	I/O	AGU	imm	ROM addr

Although the instruction fields independently steer the ASIP components, the format is not horizontal, but incorporates *local encoding*: As an example, consider the behavioral specification (table 2.1) that was provided for module "ADD".

According to definition 2.1.1, the level of abstraction in this specification is beyond the register transfer level. Module "ADD" comprises several registers, and the corresponding instruction field (bits 28:26) *indirectly* steer these registers, as well as further "hidden" RTL modules via a *local decoder*. Local decoders translate instruction fields into separate control signals for RTL modules. We call such a hardware description a *mixed-level description*, which consists of interconnected *mixed-level modules*. While the MIMOLA language also permits mixed-level descriptions, MSSQ only accepts RTL netlists. Therefore, it is necessary to break the original mixed-level description down to the RT level (fig. 2.5) by manually filling the mixed-level modules with RTL structure and

instruction bits (28:26)	"ADD" operation
000	NOP
100	s1 := s1 + m
101	s2 := s1 + m
001	sum := s1 + m
110	s1 := s2 + m
111	s2 := s2 + m
011	sum := s2 + m

Table 2.1 Specification of module "ADD" of DBB ASIP

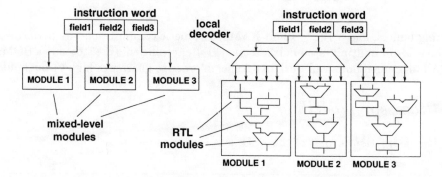

Figure 2.5 Mixed-level vs. RT-level hardware description. Arrows denote control lines.

specifying the local decoders. A MIMOLA description of a local decoder for one mixed-level module is shown in fig. 2.6. This decoder maps a 4-bit instruction field to a 9-bit "local instruction". Its output port is bit-wise connected to the control ports of RTL modules within the mixed-level module. The complete HDL model of the DBB ASIP consists of approximately 500 lines of code.

Programming

The DBB algorithm was originally described in SILAGE, i.e. in a data-flow-oriented style. In contrast, MIMOLA is an imperative language, so that the SILAGE description needed to be sequentialized. Furthermore, MSSQ has no notion of delay lines or ring buffers, so that these need to be explicitly modelled by the programmer, for which several alternatives exist in turn. We explore

```
MODULE LocalDecoder (IN instr_field: (3:0); OUT local_instr: (8:0));
BEHAVIOR IS
 CONBEGIN
  local_instruction <- CASE instr_field OF
                      %0000: %0000x0xxx;
                      %0001: %0100x0x1x;
                      %0010: %0001x0xxx;
                      %0100: %0010x0xxx;
                      %1000: %0000111x1;
                      %1001: %0000110x1;
                      %1010: %010000x0x;
                      %1011: %000011xx0;
                      %1101: %0010110x1;
                      %1110: %011000x0x;
                      %1111: %1000x1xxx;
                      ELSE : %xxxxxxxxx;
                    END;
 CONEND;
```

Figure 2.6 A local decoder in MIMOLA

some parts of the resulting implementation space by compiling three programs, each having a different level of abstraction,

High level: In the first program, the DBB algorithm is described at the pure PASCAL level, without any particular reference to the underlying hardware. Signals are mapped to variables, and delay lines are simulated by sequential data moves at the end of each sample period. Code selection, register allocation, and scheduling are completely left to the compiler.

Medium level: In the second program, storage layout is manually predefined by storage binding of variables. This also includes explicit storage allocation for ring buffers and access to ring buffer elements by means of the modulo addressing capabilities of the AGU. Code selection, (temporary) register allocation, and scheduling are left to the compiler.

Low level: In the third program, very few freedom is left to the compiler. The source program already closely reflects the machine program by predefining a sequence of control steps. Each control step comprises a set of parallel RTs. All variables are replaced by storage and register references. The compiler is

programming style	# source code statements	# machine instructions	CPU seconds (SPARC-20)
high level	84	197	602
medium level	98	98	200
low level	329	64	7

Table 2.2 Results of first application study

only responsible for binding operators to hardware resources and translating the RTs into binary machine instructions.

Results

The compilation results shown in table 2.2 clearly indicate the trade-off between code quality and manual effort for program specification. In the high-level program, the DBB signal flow graph was directly translated into a PASCAL program consisting of 84 statements. The resulting code quality is poor, however. About 50 % of the 197 generated machine instructions are due to modelling of delay lines by consecutive data moves in RAM. This problem is circumvented in the medium-level program: By modelling delay lines close to the underlying hardware, the number of instructions is halved, while the number of source statements is approximately the same as in the high-level programs. Prescribing memory layout and addressing manually, however, required significantly more programming effort. The highest code quality is achieved in the low-level program. Through extensive manual analysis of hardware capabilities, a very high resource utilization is ensured, resulting in only 64 machine instructions. The effort for writing the third program, however, comes close to assembly programming.

Although not being of primary interest, it is also worth mentioning the compilation times (table 2.2, column 4). The high-level program requires more than ten minutes of CPU time, while the low-level one is compiled within seven seconds. This is due to the fact that the runtime for MSSQ's allocation phase grows exponentially in the complexity of assignments, which in turn grows with the program abstraction level.

2.3.2 A standard DSP

The second application study deals with a complex standard DSP: the Texas Instruments TMS320C25 (cf. section 1.4.2). We focus on the problem of modelling the RTL structure of this processor for code generation with MSSQ. The information provided in the User's Manual [TI90] is behavioral, i.e. the processor is described instruction by instruction. In contrast, information about the RTL structure is rather coarse and hides many of the RTL modules. In particular, no information about the internal structure of the controller is available. This results in a similar problem as in case of the DBB ASIP: An instruction decoder has to be described, which translates TMS320C25 machine instructions into control codes for its RTL modules. In contrast to the ASIP, the TMS320C25 has a strongly encoded 16-bit instruction format which hardly shows any separation into instruction fields. Furthermore, it contains many more (in our model: 112) RTL modules, which are steered by a total of 152 separate control lines. For 114 machine instructions[2] this results in a 16×152 decoder with 114 possible outputs, i.e. 17,328 bits to be specified. Obviously, correct manual bit-wise specification of such a large decoder is nearly impossible. We therefore use a trick, which resembles a well-known compiler construction technique.

Processor modelling by bootstrapping

In compiler construction, the term *bootstrapping* denotes a procedure to *enhance* a compiler by using of the compiler itself and possibly other compilers [ASU86]. Examples are: porting the compiler to another host, optimizing it, or extending the source language that it compiles. Our bootstrapping procedure does not exactly correspond to bootstrapping in compiler construction, since it does not involve compiling a compiler. However, it realizes a similar concept: We exploit the capabilities of MSSQ in order to obtain a processor model, on which MSSQ can be applied again, so as to generate the desired machine code for the TMS320C25 processor. From the structural information provided in the TMS320C25 User's Manual, it is possible to develop an RTL data-path model together with a controller having a horizontal instruction format. MSSQ can compile source programs into microcode for this model. Thus, it can also compile a very particular source program into microcode, namely a TMS320C25 *instruction-set specification*. The resulting microcode then provides exactly the information needed for the decoder. By extending the RTL model with this decoder, we obtain a machine model, which MSSQ can use to map MIMOLA

---

[2] The TMS320C25 has a total of 133 machine instructions, of which 19 have not been modelled. These include block-move operations as well as some special instructions for memory configuration.

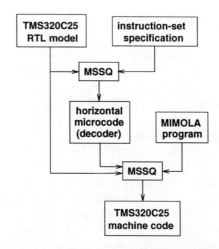

Figure 2.7 Bootstrapping procedure for the TMS320C25

program	# source code statements	# machine instructions	CPU seconds (SPARC-20)
test1	39	98	151
test2	7	22	53
test3	17	49	77
ellip	74	184	186
gcd	8	18	20
pidctrl	34	75	74
diffeq	59	99	117

Table 2.3 Results of second application study

programs into correct 16-bit instructions for the TMS320C25. This idea is visualized in fig. 2.7. Based on the processor model obtained by bootstrapping, MSSQ can compile programs into TMS320C25 machine instructions.

Results

We have used MSSQ to compile seven programs into TMS320C25 code. These include pure test programs as well as DSP algorithms. The results are listed in table 2.3. The generated machine programs in most cases suffer from unsatisfactory code quality, which is mainly due to the fact, that MSSQ allocates only statement-by-statement, but does not perform data-flow analysis between statements. This often results in redundant data move instructions. The statement-wise allocation mechanism also prevents MSSQ from exploiting potential instruction-level parallelism.

2.3.3 Conclusions from application studies

Perhaps the most important result of the above application studies is the fact, that code could be generated for two completely different DSPs by the same compiler. The reason is that MSSQ uses detailed HDL processor models. Machine code is directly derived from the hardware model. This concept is also well-suited for machine-independent DSP code generation. Therefore, we adopt this approach to retargetable compilation also for the technique described in the next chapter.

However, as became obvious in the application studies, pure RTL models are sometimes not a convenient way of describing processors. Generation of RTL models from higher-level models is possible, but requires knowledge about internal compiler techniques. Thereby, the degree of retargetability is reduced from "machine-independent" to "portable". Furthermore, the greedy code selection and register allocation algorithms used in MSSQ cause insufficient code quality. The same holds for the MSSQ's heuristic code compaction technique, which works well for horizontal instruction formats, but fails to exploit potential parallelism in case of strongly encoded formats. In addition, DSP-specific hardware features like mode registers or address generation units should be recognized by the compiler.

Solutions to these problems are presented in the remainder of this book. In the next chapter we revisit the problem of processor modelling at different levels of abstraction. We present a powerful and user-friendly processor model analysis procedure, which exploits research results from logic synthesis.

3
INSTRUCTION-SET EXTRACTION

Extraction is a very common concept in ECAD, which is generally used as a means of validation, such as *parameter extraction* or *netlist extraction* from mask layout. In this chapter we present an extraction technique intended to provide the necessary link between hardware-oriented processor models and advanced code generation algorithms: Instruction-set extraction reads an HDL processor model and emits the set of valid instructions for the specified processor. The organization of this chapter is as follows. First, we analyze different processor description styles and we define the necessary terminology. The main part of this chapter is constituted by algorithms for extracting the instruction set from an HDL processor model. The overall extraction process consists of three main phases: construction of an internal processor model, behavioral analysis of single modules, and structural analysis for composing local module operations to global instruction patterns.

3.1 PROCESSOR DESCRIPTION STYLES

In the area of compiler construction, mostly behavioral processor descriptions are used. A behavioral description consists of a list of opcodes and mnemonics for machine instructions and specifies the behavior of each instruction. For processors with instruction-level parallelism such a description is, however, often uncomfortable, because there is no clear interface by means of "instructions". Instead, an instruction is a collection of entities of finer granularity, namely register transfers. This problem is reflected by the fact, that instruction-set descriptions in User's Manuals of standard DSPs are less clear to read than descriptions of CISCs and RICSs, because of many footnotes and cross-references describing possible combinations of RTs and exceptions. For processors with a more horizontal instruction format the instruction sets are therefore better described hierarchically: The first level lists the available *instruction types*, e.g. ALU or move operations, while the second level specifies the possible operands and destinations. In case of a purely horizontal instruction format, the processor is most naturally described as a netlist of RTL components (e.g. example processor from section 2.1). In contrast to behavioral descriptions, RTL netlists *implicitly* represent all available RTs and valid combinations of the same. Furthermore, RTL netlists are also very common models in ECAD systems. In fact, the best processor description style can only be selected by the user for each particular case, mainly dependent on three factors:

- The available **documentation**: Although the different styles can be transformed into each other, it is in general most efficient to directly use the documentation that is provided with the processor, e.g. an instruction-set description in a User's Manual or a schematic generated by an ECAD tool.

- The degree of **instruction-level parallelism**: The more parallelism is present in an instruction format, the finer the granularity of the description should be in order to avoid too many "footnotes".

- The **processor type**: In case of standard DSPs with a fixed architecture, the description should be as concise as possible and hide unnecessary details. In case of an ASIP, whose architecture is not fixed in advance, a more detailed description facilitates customization of the ASIP hardware towards a certain application.

With respect to MIMOLA processor models, we distinguish three categories of *description styles*:

3.1.1 Definition
A MIMOLA processor model is called **behavioral**, if it consists of only one behavioral module. Otherwise, if the model consists of structural modules and behavioral modules, and all behavioral modules are RTL modules, then it is called **structural**. All other models are called **mixed-level** models.

Abstracting from the description language, all previous approaches to (retargetable) embedded code generation are based on exactly one of those three styles: Behavioral models are used by Rimey/Hilfinger [RiHi88] and in CBC

Instruction-set extraction

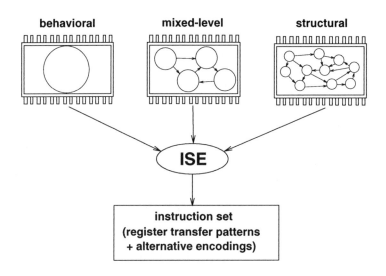

Figure 3.1 Functionality of instruction-set extraction

[FaKn93], structural models are the basis of MSSQ [Nowa87a], and mixed-level models are preferred by Wess [Wess92], Philips [SMT+95], and in CodeSyn [LMP94a] and CHESS [FVM95]. However, no attempt has been made so far to leave selection of a certain style to the user. This leads to the concept of *instruction-set extraction* (ISE), as depicted in fig. 3.1. ISE is a frontend, which accepts MIMOLA processor models in any of the above styles, and which yields the set of instructions available on the processor in form of register transfer patterns, as well as the corresponding partial instructions (or binary encodings). The main advantages of ISE are the following:

Versatility: The choice of a certain processor description style is not prescribed by the compiler, but is completely left to the user. Dependent on the intended application and the available documentation, either more abstract or more hardware-oriented processor models can be used.

Uniformity: Most previous approaches use very tool-specific processor description formats. ISE is based on a well-tried and well-documented hardware description language which provides a close link between the compiler and ECAD tools. The output of ISE is always an instruction set, independent of the granularity of the input model. Thus, code generation algorithms can operate on a uniform representation of the processor.

Efficiency: Code generation based on graph models like in MSSQ or CHESS inherently involves a high degree of redundancy: For each occurrence of a certain operation in the source program, traversal of the graph model is unnecessarily repeated. By determination of valid RT patterns and checking for conflicts only once in advance, this redundancy is avoided, resulting in lower runtime requirements for code generation.

Earlier approaches to instruction-set extraction ssuffer from the restriction, that too many a priori assumptions are made about the processor instruction format and component types. For instance, the approaches in [BBM86, HeGl94] assume a horizontal instruction format. Monahan and Brewer's tool [MoBr95] takes instruction conflicts into account, but only accepts structural models composed of four basic component types. As we argue in the following section, presence of encoded instruction formats, mode registers, and complex components demands for a more thorough model analysis phase.

3.2 ANALYSIS OF CONTROL SIGNALS

As explained in section 2.1, the behavior of MIMOLA modules is specified by assignments, either variable or signal assignments. The behavior of a module is either *fixed*, i.e. its output is only dependent on *data* that are being processed, or it is *programmable*, i.e. it is dependent both on data and the *instructions* that are executed. In case of programmable behavior, the functionality of a module is controlled by distinguished input signals, which we call *control signals*. These control signals are received by modules via *control ports*. For microprogrammable processors, the compiler can adjust control signals by appropriate setting of microinstruction bits. By *control signal analysis*, we denote the process of identifying the necessary control signal settings for each assignment in a module, and adjusting these control signals through tracing back the path from control ports to the instruction memory in a hardware structure. This also implies checking for conflicts between control signals in order to reveal invalid operations or combinations of the same.

Code generators which are not based on structural models do not perform control signal analysis at all, but assume that this task has already been performed in advance. This is, however, rather inconvenient, if the processor is only described as a netlist, which is often the case for ASIPs. A slight change in the RTL model may have a large impact on the overall instruction set. Therefore, "non-structural" approaches do not well support customization of

Instruction-set extraction

```
MODULE PipelinedALU (IN buffer_output, add, update_inputs: Bit;
                     IN i1, i2: Word;
                     OUT outp: Word);
BEHAVIOR IS
 VAR R1, R2, R3: Word;
 CONBEGIN
  CASE buffer_output OF
   1: CONBEGIN
       R3 := (IF add THEN R1 + R2 ELSE R1 - R2);
       outp <- R3;
      CONEND;
   0: outp <- (IF add THEN R1 + R2 ELSE R1 - R2);
  END;
  IF update_inputs THEN
   CONBEGIN
    R1 := i1;
    R2 := i2;
   CONEND;
 CONEND;
```

Figure 3.2 Pipelined ALU module

ASIPs. In contrast, code generators based on structural models, such as previous MIMOLA-based compilers, suffer from the restriction, that only pure RTL models are accepted.

3.2.1 Complex modules

While RTL specification of simple modules like registers is not a severe problem, the situation becomes much worse for complex (mixed-level) modules with nested **IF** and **CASE** constructs, such as the pipelined ALU in fig. 3.2. In order to eliminate the restriction of handling RTL models only we use a fine-grained approach for control signal analysis, based on single assignments within a module description instead of modules themselves. Consider module **PipelinedALU** again. We can transform its behavioral specification into a tabular form, which lists all possible assignments and the according control signals (table 3.1). For instance, the assignment "outp <- R1 + R2" is executed, exactly if buffer_output = 0 and add = 1, while the value of update_inputs is don't care in this case. Such a "truth table" format expresses the behavior of programmable modules independently of the syntactical structure of its

assignment	buffer_output	add	update_inputs
R3 := R1 + R2	1	1	x
R3 := R1 - R2	1	0	x
outp <- R3	1	x	x
outp <- R1 + R2	0	1	x
outp <- R1 - R2	0	0	x
R1 := i1	x	x	1
R2 := i2	x	x	1

Table 3.1 Possible assignments in module PipelinedALU

```
MODULE Decoder5x1 (IN inp: (4:0);
                   OUT outp: Bit);
BEHAVIOR IS
CONBEGIN
  outp <- CASE inp OF
    %00x0: 1;
    %0001: 0;
    %1x11: 1;
    %0x11: 0;
    %110x: 1;
  END;
CONEND;
```

```
MODULE Decoder5x1 (IN inp: (4:0);
                   OUT outp: Bit);
BEHAVIOR IS
CONBEGIN
  outp <- (NOT (inp.(0) OR inp.(2) OR inp.(3)))
          OR
          (inp.(3) AND ((NOT(inp.(1)) AND inp.(2))
                        OR (inp.(0) AND inp.(1))))
CONEND;
```

a)   b)

Figure 3.3 Decoder module: a) in RT-level form, b) in gate-level form.

description.

In some applications, it may also be desirable to partially extend the model abstraction level down to the gate level. Consider the decoder in fig. 3.3 a). An equivalent gate-level description is shown in fig. 3.3 b), which could have been generated by a logic synthesis tool. Analysis of gate-level descriptions requires knowledge of the semantics of logical operators. This observation as well as the above "truth table" format indicate a resonable approach to control signal analysis: If control signals are interpreted as *Boolean variables*, then analysis can be performed in a uniform way by means of *Boolean functions*, i.e. mappings

$$f : \{0,1\}^n \to \{0,1\}, \quad n \in \mathbb{N}.$$

Using Boolean functions as a *data type* in a program demands for a powerful data structure. Such a data structure exists in form of *binary decision diagrams*, which are briefly presented in the following section.

3.3 BINARY DECISION DIAGRAMS

Well-known representations of Boolean functions are truth tables, circuits, and Boolean formulas. The applicability of a certain representation is mainly determined by the set of operations that should be efficiently supported. Important operations include evaluation, synthesis, equality test, and substitution of variables.

Unfortunately, almost all of the 2^{2^n} different n-ary Boolean functions have exponential size in the above representations, which in turn excludes efficient manipulation of those functions. Nevertheless, some representations are better than others, because they make use of the structure of Boolean functions. This structure is exploited in *binary decision diagrams* (BDDs). BDDs are DAGs, consisting of inner nodes (including the BDD root) representing Boolean variables, and leaves representing the values 0 and 1. Each inner node has two outgoing edges labelled with 0 and 1, respectively. Starting at the root of a BDD and traversing it down to a leaf, according to a given input vector, means to evaluate a Boolean function. A BDD for the function

$$f(x_1, x_2, x_3) \quad = \quad \overline{x_1} \cdot x_2 \cdot x_3 \quad \vee \quad x_1 \cdot \overline{x_2} \cdot x_3 \quad \vee \quad x_1 \cdot x_2 \cdot x_3$$

is shown in fig. 3.4 a). The concept of BDDs has been developed already some decades ago, but a breakthrough has been achieved only in 1985 by Bryant [Brya85, Brya86], who proposed *reduced ordered* BDDs (ROBDDs). For these, a total ordering of the Boolean variables is defined, which has to be obeyed on each directed path between inner BDD nodes. For instance, the BDD in fig. 3.4 a) obeys the ordering $x_1 < x_2 < x_3$. Furthermore, ordering and semantic preserving *transformation rules* [Brya92] are applied, so as to remove redundant subgraphs. The ROBDD for the above Boolean function is shown in fig. 3.4 b). Two features make ROBDDs *the* state-of-the-art representation of Boolean functions:

- ROBDDs are **canonical** for a fixed variable ordering. This facilitates the equivalence test and its important special cases.

- ROBDDs have **polynomial size** (in the number of variables) for many more practical Boolean functions than other representations, and **polyno-**

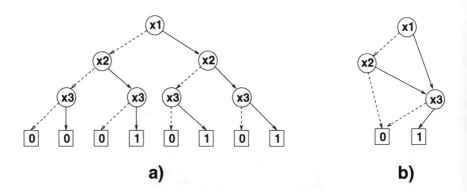

Figure 3.4 Binary decision diagrams: a) a general BDD, b) equivalent reduced ordered BDD. Solid arrows denote 1-edges, dashed arrows denote 0-edges

mial-time algorithms (in the BDD size) are available for all important operations on ROBDDs.

In this book, we use ROBDDs (or BDDs, for short) mainly as a *black box* that provides efficient manipulation of Boolean functions. Boolean functions form the basis of the instruction-set model presented in the following.

3.4 INSTRUCTION-SET MODEL

For machines with instruction-level parallelism, an instruction is a set of parallel RTs, for each of which a set of alternative encodings may be available. Complete enumeration of all available instructions on a highly parallel machine may lead to an instruction-set model of size exponential in the instruction word-length. Furthermore, generation of assembly code and binary code must not be separated, because selection of a good encoding for a register transfer is context-dependent. Our instruction-set model therefore uses register transfers as the primitive entities instead of complete instructions, and maintains all available encodings for each RT.

3.4.1 Register transfer patterns

We use the term register transfer *patterns* in order to distinguish these objects from register transfers. A register transfer pattern (RTP) is a *template* for

Instruction-set extraction

$MIM(P)$:	MIMOLA model of a target processor P
REG:	The set of all scalar module variables (registers) in $MIM(P)$
MEM:	The set of all non-scalar module variables (memories) in $MIM(P)$
P_{IN}:	The set of primary input ports in $MIM(P)$
P_{OUT}:	The set of primary output ports in $MIM(P)$
$read(r)$	read access to register $r \in REG$
$read(m,a)$	read access to memory cell $m[a], m \in MEM$
$read(p)$	read access to primary input port $p \in P_{IN}$
$write(r)$	write access to register $r \in REG$
$write(m,a)$	write access to memory cell $m[a], m \in MEM$
$write(p)$	write access to primary output port $p \in P_{OUT}$

Table 3.2 Notations for definition of RT patterns

a single-cycle RT on a microprogrammable processor, and RTs are instances of RTPs. A register transfer reads input values from registers, memory cells, and/or input ports, performs a computation, and assigns the result to a destination. In this way, RTPs reflect the data-path capabilities of a target processor. An example will be shown in section 3.9. For definition of RTPs, we refer to the notations from table 3.2.

3.4.1.1 Definition
An **RT expression** is one of the following items:

- a **binary constant** $B \in \{0, 1, x\}^+$

- a **scalar read access** $read(r)$ or $read(p)$

- an **indexed read access** $read(m, a)$, where "address" a is an RT expression

- a **complex expression** $op(e_1, \ldots, e_k)$, where op is a predefined MIMOLA operator with arity $k \geq 1$, and $e_1, \ldots, e_k$ are RT expressions

- a **subrange expression** $e' = e.(hi : lo)$, where e is an RT expression and $(hi : lo)$ denotes a bit index subrange

An **RT destination** is either

- a **scalar write access** $write(r)$ or $write(p)$, or
- an **indexed write access** $write(m, a)$, where a is an RT expression

A **register transfer pattern** is a pair $\mathcal{R} = (d, e)$, where d is an RT destination and e is an RT expression.

3.4.2 Register transfer conditions

From a hardware-oriented point of view, register transfers take place during execution of a machine program, if a certain *machine state* is present. Execution of an RT in general depends on some *static* conditions, i.e. a certain setting of the current instruction and a certain state of mode registers. All decisions concerning static conditions are already made at *compile time*, because the code generator is responsible for generating appropriate instruction settings and mode register states. In addition, execution of an RT may also depend on whether or not a *dynamic (run time)* condition is fulfilled, e.g. equality of two register contents. We can therefore associate a *register transfer condition* (RTC) with each RT. An RT is executed if and only if its RTC evaluates to *TRUE* for a given machine state. With the concept of RTCs, we can consider a processor as a machine, which in every control step executes the same set of parallel *guarded operations*, each of which has the form

```
IF <register transfer condition> THEN <do register transfer>
```

In general, RTCs may be arbitrary combinations of static and dynamic conditions, and therefore may be considerably complex. We use a canonical, bit-wise representation of RTCs, which covers static and dynamic conditions in a uniform way.

3.4.2.1 Definition
Let L_w denote the instruction word-length of a processor. The **instruction bit variables** are defined as the set of Boolean variables

$$IBV = \{I_k \mid k \in \{0, \ldots, L_w - 1\}\}$$

Instruction-set extraction

Let $MR \subseteq REG$ be the (possibly empty) set of mode registers[1] of a processor, and let $wid(mr)$ denote the bit-width of $mr \in MR$. The **mode register variables** are defined as the set of Boolean variables

$$MRV = \bigcup_{mr \in MR} \{M_{mr,k} \mid k \in \{0, \ldots, wid(mr) - 1\}\}$$

Let $COMP$ denote the set of all *comparisons* permissible on a processor, i.e. the set of RT expressions of the form $op(e, e')$, where $op \in \{=, \neq, <, >, \leq, \geq\}$ is a comparison operator. The **dynamic condition variables** are defined as the set of Boolean variables

$$DCV = \{D_c \mid c \in COMP\}$$

A **register transfer condition** is a Boolean function

$$F : \{0, 1\}^K \rightarrow \{0, 1\}$$

represented by a BDD on the set of Boolean variables

$$IBV \cup MRV \cup DCV$$

with

$$K = |IBV \cup MRV \cup DCV|$$

Having defined register transfer patterns and conditions, we can specify the meaning of *instruction-set extraction*.

3.4.2.2 Definition
A **guarded register transfer pattern** (GRTP) is a pair $(\mathcal{R}, F)$, where $\mathcal{R}$ is a register transfer pattern, and F is a register transfer condition.

Instruction-set extraction is the task of computing all GRTPs for a processor model $MIM(P)$.

GRTPs capture all information required for code generation: RTPs represent the RTs that a processor can execute, while RTCs represent the required machine state in terms of partial instructions and mode register states, as well as possible dynamic conditions. Using BDDs as a data structure for RTCs permits canonical representation of conditions and early detection of *intra-RT conflicts*, which is exemplified in the next section.

---
[1] We assume, that mode registers (storing control codes) and regular registers (storing data) are disjoint. Furthermore, we assume, that all mode registers are scalar, i.e. there are no "mode memories". Both assumptions are satisfied for realistic processors.

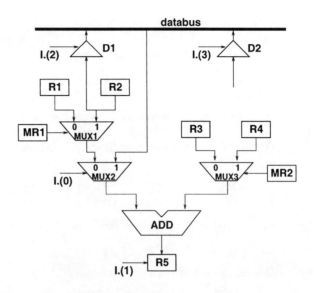

Figure 3.5 Partial RT-level hardware structure

3.4.3 Example

Consider the partial RTL structure shown in fig. 3.5. The multiplexers MUX1, MUX3 are steered by mode registers (MR1 and MR2). All other programmable modules receive control codes from the instruction word "I". Suppose that we want to analyze the RTC for RTP "R5 := R2 + R3". Routing the value of R2 to the left input of module ADD via multiplexers MUX1, MUX2 can be realized by control codes MR1 = "1" and I.(0) = 0. By setting MR2 = 0, the contents of register R3 are switched to the right input of ADD. Furthermore, loading R5 with a result from ADD can be enabled by setting I.(1) = 1. In total, the corresponding RTC[2] is

$$F_1 = \overline{I_0} \cdot I_1 \cdot M_{MR1} \cdot \overline{M_{MR2}}$$

There exists also an alternative route for transportation of R2 to ADD: If bus driver D1 is activated by setting I.(2) = 1, and MUX2 passes its right input (I.(0) = 1), then the value of R2 is routed via bus databus. A potential bus conflict needs to be avoided by setting driver D2 to a tristate mode (I.(3) = 0). The control codes I.(1) = 1 and MR2 = 0 remain unchanged. The resulting RTC is

$$F_2 = I_0 \cdot I_1 \cdot I_2 \cdot \overline{I_3} \cdot \overline{M_{MR2}}$$

---

[2] Since MR1, MR2 are 1-bit registers, we omit the bit index for mode register variables.

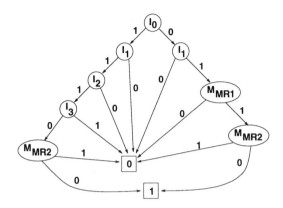

Figure 3.6 BDD representation of a register transfer condition

Since F_1 and F_2 are alternatives, we obtain the GRTP

$$(\text{R5} := \text{R2} + \text{R3}, F_1 \vee F_2)$$

For the variable ordering

$$I_0 < I_1 < I_2 < I_3 < M_{MR1} < M_{MR2}$$

$F_1 \vee F_2$ is represented by the BDD depicted in fig. 3.6. Note, that this RTC representation is completely independent (except for exchange of identifiers) of the style and syntax of a concrete MIMOLA model for the structure from fig. 3.5. That is, it does depend on whether the MIMOLA model is structural, behavioral, or mixed, and whether programmable behavior of modules is described through IF or CASE statements. The information required for code generation, namely the machine state with respect to partial instructions and mode register states, is the same in all cases.

Moreover, the canonical representation of RTCs by BDDs can prevent consideration of RTPs, which are invalid due to instruction encoding, for instance if RTL modules have *shared* control signals in order to reduce the instruction word-length. Suppose, that modules MUX2 and D2 in the example structure are steered by the same instruction bit, and modules MUX1 and MUX3 are controlled by the same mode register. The consequences for pattern R5 := R2 + R3 can be evaluated by computing the function $F' = F'_1 \vee F'_2$ with

$$F'_1 = (F_1|_{M_{MR2}=M_{MR1}})|_{I_3=I_0}, \qquad F'_2 = (F_2|_{M_{MR2}=M_{MR1}})|_{I_3=I_0}$$

so that

$$\begin{aligned} F' &= F'_1 \vee F'_2 \\ &= \overline{I_0} \cdot I_1 \cdot M_{MR1} \cdot \overline{M_{MR1}} \quad \vee \quad I_0 \cdot I_1 \cdot I_2 \cdot \overline{I_0} \cdot \overline{M_{MR1}} \\ &\equiv 0 \end{aligned}$$

Although the RTP R5 := R2 + R3 is valid with respect to the *data-path*, it has a non-satisfiable RTC. We call this situation an **intra-RT conflict**. Since the corresponding BDD is simply a single zero node, one can easily check for intra-RT conflicts already during control signal analysis. Only those GRTPs $(\mathcal{R}, F)$ are valid, for which $F \not\equiv 0$.

3.4.4 No-operations

So far we have only considered "useful" operations, i.e. register transfers. Additionally, a detailed instruction-set model for retargetable code generation must account for "no-operations" (NOPs). All general-purpose processors comprise NOP instructions, which do not perform any operation but are primarily used for "busy waiting" in a machine program, or for "padding" manually patched programs. In our model, intended for machines with instruction-level parallelism, NOPs play a completely different role, namely avoidance of undesired side effects. Therefore, NOPs must not be considered as "instructions", but (similar to RTPs) as primitive operations associated with a certain "destination". While other approaches to retargetable code generation assume, that NOPs are implicit in each machine instruction, only *explicit* handling of NOPs guarantees correct code generation for arbitrary instruction formats. As we will see later during discussion of *code compaction*, NOPs are also very important for exploiting the rather subtle type of parallelism in some DSPs.

3.4.4.1 Definition
A **no-operation** is a pair (d, F), where $d \in REG \cup MEM$ is a register or a memory, and F is a register transfer condition.

Like GRTPs, we can consider NOPs as guarded operations which are "executed" under certain conditions. The semantics of a NOP is, that the destination d retains its state for one machine cycle, if its RTC F evaluates to $TRUE$ for a given machine state.

The instruction-set model delivered by ISE is the set of all GRTPs and all NOPs. Systematic extraction of this model from a MIMOLA processor descrip-

Instruction-set extraction

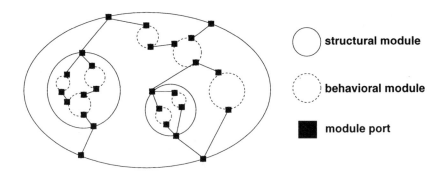

Figure 3.7 Hierarchical processor structure

tion proceeds in three sequential phases: First, an internal graph-like model of the target processor is constructed, which represents modules and interconnections. Then, ISE is performed on this graph model in a bottom-up fashion: After analyzing the behavior of modules *locally*, the hardware structure (if specified) is investigated *globally*, i.e. across module boundaries, in order to construct the set of GRTPs. From this set, also the set of NOPs can be derived. The detailed explanation of ISE is given in the next three sections.

3.5 INTERNAL PROCESSOR MODEL

In general, MIMOLA descriptions are *hierarchical structures* as shown in fig. 3.7. The inner nodes of such a structure are either in turn structural modules, or behavioral modules, which constitute the leaves of a structure. Connections between modules can be represented by directed edges between module ports. The internal processor model used for instruction-set extraction is therefore a graph model, which captures the hierarchy of processor components and interconnections in terms of wires and busses. Since construction of this internal representation from a textual MIMOLA description does not require special techniques, we just provide an example: Fig. 3.8 shows the internal graph model of the simple processor from section 2.1. In this case, a flat netlist of behavioral modules surrounded by a structural module is obtained.

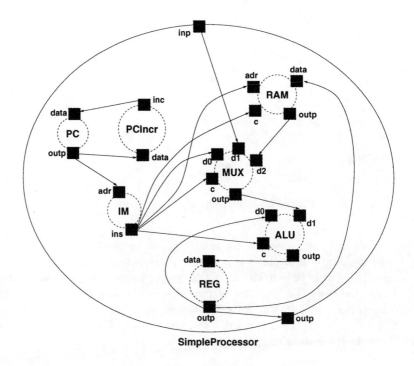

Figure 3.8 Internal representation of SimpleProcessor. Bit index subranges of connections are omitted.

3.6 BEHAVIORAL ANALYSIS

The goal of behavioral analysis is to generate a representation of behavioral modules by means of "local" GRTPs. A behavioral MIMOLA module is characterized by a set of *module ports*, a set of *module variables*, and a set of *concurrent statements*. All statements fall into one of the following categories:

Signal assignments of the form <port> <- <expression>, where <port> denotes an OUT or INOUT port, and <expression> is an RT expression or a conditional (CASE or IF) expression.

Variable assignments of the form <variable> := <expression> for scalar module variables or <variable>[<index>] := <expression> for array module variables.

IF-statements of the form
IF <condition> THEN <concurrent statements>

Instruction-set extraction

```
            ELSE <concurrent statements>
```
where `<condition>` is a 1-bit expression which evaluates to 0 or 1. The ELSE branch is optional.

CASE-statements of the form
```
   CASE <selector> OF
      <constant_1_1>,...,<constant_1_n1>:  <concurrent statements>
      ...
      <constant_m_1>,...,<constant_m_nm>:  <concurrent statements>
      ELSE: <concurrent statements>
   END
```
where `<selector>` refers to an IN port, a scalar module variable, or an index subrange of the same. If `<selector>` matches one of the values `<constant_i_1>,...,<constant_i_ni>`, then the corresponding set of concurrent statements is executed. Otherwise, the statements in the ELSE branch are executed.

TRISTATE statements of the form `TRISTATE <port>`, where `<port>` denotes an OUT or INOUT port.

3.6.1 Definitions

The representation of behavioral modules we want to obtain is defined in the following:

3.6.1.1 Definition

Let B be a behavioral module with input ports $p_1, \ldots, p_n$ and scalar module variables $v_1, \ldots, v_m$.

The **local port variables** for B are defined as the set of Boolean variables

$$PV(B) = \bigcup_{i=1}^{n} \{P_{p_i,k} \mid k \in \{0, \ldots, wid(p_i) - 1\}\}$$

The **local mode register variables** for B are defined as the set of Boolean variables

$$MRV(B) = \bigcup_{j=1}^{m} \{M_{v_j,k} \mid k \in \{0, \ldots, wid(v_j) - 1\}\}$$

Let $COMP(B)$ denote the set of all *comparisons* permissible in module B, i.e. the set of RT expressions of the form $op(e, e')$, where $op \in \{=, \neq, <, >, \leq, \geq\}$ is a comparison operator. The **local dynamic condition variables** for B are defined as the set of Boolean variables

$$DCV(B) = \{D_c \mid c \in COMP(B)\}$$

A **local condition** for B is a Boolean function

$$F_B : \{0,1\}^{K_B} \to \{0,1\}$$

represented by a BDD on the set of Boolean variables

$$PV(B) \cup MRV(B) \cup DCV(B)$$

with

$$K_B = |PV(B) \cup MRV(B) \cup DCV(B)|$$

The idea underlying these definitions is, that all operations within a module are *locally* steered by control lines originating from module input ports or from scalar module variables, or that these operations depend on (dynamically evaluated) signal comparisons. Therefore, the conditions for local operations can be interpreted as certain settings of the above Boolean variables. The way in which local operations depend on these variables is encoded in (possibly nested) IF and CASE statements/expressions in the module description. A canonical, syntax-independent representation of the module behavior can be achieved by eliminating all IF and CASE constructs, and associating a guard with each assignment, similar to GRTPs. In this way, one obtains a tabular representation as already introduced in section 3.2.1. The following types of guarded assignments may occur.

3.6.1.2 Definition
A **guarded port assignment** for B is a tuple (p, E, F_B), where p is an OUT or INOUT port of B, E is an RT expression, and F_B is a local condition.

A **guarded tristate assignment** for B is a tuple $(p, TRISTATE, F_B)$, where p is an OUT or INOUT port of B, and F_B is a local condition.

A **guarded variable assignment** for B is a tuple (v, E, F_B), where v is a scalar module variable of B, E is an RT expression, and F_B is a local condition. For array module variables, guarded variable assignments have the form (v, I, E, F_B), where the "index" I is an RT expression as well.

```
(1)  algorithm EXTRACTGUARDEDASSIGNMENTS
(2)  input: set S of concurrent statements
(3)  output: set GA of all guarded assignments for S
(4)  var ga: set of guarded assignments;
(5)  begin
(6)     GA := ∅;
(7)     for all s ∈ S do
(8)        case s of
(9)           SIGNAL ASSIGNMENT:
(10)             ga := EXTRACTSIGNALASSIGNMENT(s);
(11)          VARIABLE ASSIGNMENT:
(12)             ga := EXTRACTVARIABLEASSIGNMENT(s);
(13)          IF-STATEMENT:
(14)             ga := EXTRACTIFSTATEMENT(s);
(15)          CASE-STATEMENT:
(16)             ga := EXTRACTCASESTATEMENT(s);
(17)          TRISTATE-STATEMENT:
(18)             ga := EXTRACTTRISTATE(s);
(19)       end case
(20)       GA := GA ∪ ga;
(21)    end for
(22)    return SIMPLIFY(GA);
(23) end algorithm
```

Figure 3.9 Extraction of guarded assignments from MIMOLA statements

3.6.2 Extraction algorithms

Extraction of all guarded assignments for a behavioral module is performed by calling the algorithm shown in fig. 3.9 for the set of concurrent statements in the module description. For each statement, an extraction subroutine corresponding to the statement type is invoked. Subroutine SIMPLIFY finally merges those guarded assignments, which only differ in their local conditions.

We consider extraction in case of IF-statements (fig. 3.10): First, the sets of guarded assignments are separately (and for nested conditions recursively) extracted from the THEN and ELSE branches of the IF-statement (lines 11-12). Then, the IF-condition is transformed into a local condition F (line 14).

---

(1) **algorithm** EXTRACTIFSTATEMENT
(2) **input:** IF-statement s of the form
(3)   IF <condition>
(4)   THEN <concurrent statement set S_1>
(5)   ELSE <concurrent statement set S_2>
(6) **output:** set GA of all guarded assignments for s
(7) **var** GA_{THEN}, GA_{ELSE}: **set of** guarded assignments;
(8) ga: guarded assignment;
(9) F, F_{ga}: Boolean function;
(10) **begin**
(11) $GA_{THEN} :=$ EXTRACTGUARDEDASSIGNMENTS(S_1);
(12) $GA_{ELSE} :=$ EXTRACTGUARDEDASSIGNMENTS(S_2);
(13)   /* $GA_{ELSE} := \emptyset$, if no ELSE-branch is present in s */
(14) $F :=$ MAKECONDFUNC(<condition>);
(15)   **for all** $ga \in GA_{THEN}$ **do**
(16) $F_{ga} :=$ local condition of ga;
(17) $F_{ga} := F_{ga} \wedge F$;
(18)   **end for**
(19)   **for all** $ga \in GA_{ELSE}$ **do**
(20) $F_{ga} :=$ local condition of ga;
(21) $F_{ga} := F_{ga} \wedge \overline{F}$;
(22)   **end for**
(23) $GA := GA_{THEN} \cup GA_{ELSE}$;
(24)   **return** SIMPLIFY(GA);
(25) **end algorithm**

Figure 3.10 Extraction of guarded assignments from IF-statement

There, we make use of subroutine

$$\text{MAKECONDFUNC} : \text{<condition>} \mapsto (F : \{0,1\}^{K_B} \to \{0,1\})$$

which maps an IF-condition into a local condition. MAKECONDFUNC is defined as follows:

- If <condition> is a 1-bit index subrange "$v.(k)$" of module variable v, then MAKECONDFUNC yields a Boolean function F, with $F = M_{v,k}$ ($M_{v,k}$ is a local mode register variable).

Instruction-set extraction

- If `<condition>` is a 1-bit index subrange "$p.(k)$" of module input port p, then MAKECONDFUNC yields a Boolean function F, with $F = P_{p,k}$ ($P_{p,k}$ is a local port variable).

- If `<condition>` is a comparison $c = op(e_1, e_2)$ ($op \in \{=, \neq, <, >, \leq, \geq\}$), then MAKECONDFUNC yields a Boolean function F, with $F = D_c$ (D_c is a local dynamic condition variable).

- If `<condition>` is a logical expression $op(e_1, e_2)$, ($op \in \{AND, OR, \ldots\}$), then MAKECONDFUNC yields a Boolean function $F = op(F_1, F_2)$, with $F_1 =$ MAKECONDFUNC(e_1) and $F_2 =$ MAKECONDFUNC(e_2). Special cases are unary expressions of the form $NOT(e)$, for which the Boolean function $\overline{F}$, with $F =$ MAKECONDFUNC(e), is computed.

- All other cases are treated as an "error", i.e. control signals must not originate from any other than the above sources. In case of an error, the processor model needs to be revised.

The total set of guarded assignments is finally obtained by combining the guarded assignments from the THEN and ELSE branches with F and $\overline{F}$, respectively (lines 15-22).

The algorithm for extraction from CASE-statements is slightly more complex, because a number of different conditions have to be taken into account. The extraction algorithm in fig. 3.11 works as follows: For each possible value N_{ij} of the CASE selector, a "representative" Boolean function F_{ij} is constructed (lines 15-19). This is performed by subroutine

MAKESELECTORFUNC : `<selector>` $\times \{0, 1, x\}^+ \mapsto (F_{ij} : \{0, 1\}^{K_B} \to \{0, 1\})$

where `<selector>` is a subrange expression "$p.(hi : lo)$" for a module input port p, a subrange expression "$v.(hi : lo)$" for a module variable v, or a logical expression. Let $(n_{ij}^1, \ldots, n_{ij}^s)$, with $n_{ij}^k \in \{0, 1, x\}$, be the binary representation of N_{ij} with width $s = hi - lo + 1$. The Boolean function F_{ij} delivered by MAKESELECTORFUNC is defined as

$$F_{ij} = Z_{ij}^1 \wedge \ldots \wedge Z_{ij}^s$$

(1) **algorithm** EXTRACTCASESTATEMENT
(2) **input:** CASE-statement s of the form
(3)   CASE <selector> OF /* $N_{ij} \in \{0,1,x\}^+$ */
(4) $N_{11}, \ldots, N_{1k_1}$:   <concurrent statement set S_1>
(5) $\ldots$
(6) $N_{m1}, \ldots, N_{mk_m}$:   <concurrent statement set S_m>
(7)   ELSE: <concurrent statement set S_{m+1}>
(8)   END
(9) **output:** set GA of all guarded assignments for s
(10) **var** L: Boolean function;
(11) ga: guarded assignment;
(12) F, G: **array of** Boolean function;
(13) GA: **array of** set of guarded assignments;
(14) **begin**
(15)   **for all** $i \in \{1, \ldots, m\}$ **do**
(16)   **for all** $j \in \{1, \ldots, k_i\}$ **do**
(17) $F_{ij} :=$ MAKESELECTORFUNC(<selector>, N_{ij});
(18)   **end for**
(19)   **end for**
(20)   **for all** $i \in \{1, \ldots, m\}$ **do**
(21) $G_i := (\bigvee_{j=1}^{k_i} F_{ij})$;
(22) $GA_i :=$ EXTRACTGUARDEDASSIGNMENTS(S_i);
(23)   **end for**
(24) $GA_{m+1} :=$ EXTRACTGUARDEDASSIGNMENTS(S_{m+1});
(25)   **for all** $i \in \{1, \ldots, m+1\}$ **do**
(26)   **for all** $ga \in GA_i$ **do**
(27) $L :=$ local condition of ga;
(28)   **if** $i < m+1$ **then** $L := L \wedge G_i$ **else** $L := L \wedge (\overline{\bigvee_{j=1}^{m} G_i})$;
(29)   **end if**
(30)   **end for**
(31)   **end for**
(32)   **return** SIMPLIFY($\bigcup_{i=1}^{m+1} GA_i$);
(33) **end algorithm**

Figure 3.11 Extraction of guarded assignments from CASE-statement

where for all $k \in \{1, \ldots, s\}$

$$Z_{ij}^k = \begin{cases} P_{p,k} & \text{if } \texttt{<selector>} \text{ refers to port } p \\ & \text{and } n_{ij}^k = 1 \\ \overline{P_{p,k}} & \text{if } \texttt{<selector>} \text{ refers to port } p \\ & \text{and } n_{ij}^k = 0 \\ M_{v,k} & \text{if } \texttt{<selector>} \text{ refers to variable } v \\ & \text{and } n_{ij}^k = 1 \\ \overline{M_{v,k}} & \text{if } \texttt{<selector>} \text{ refers to variable } v \\ & \text{and } n_{ij}^k = 0 \\ \textsc{MakeCondFunc}(e) & \text{if } \texttt{<selector>} \text{ is a logical expr. } e \\ & \text{and } n_{ij}^k = 1 \\ NOT(\textsc{MakeCondFunc}(e)) & \text{if } \texttt{<selector>} \text{ is a logical expr. } e \\ & \text{and } n_{ij}^k = 0 \\ 1 & \text{if } n_{ij}^k = x \quad \text{(don't care)} \end{cases}$$

so that F_{ij} evaluates to $TRUE$ exactly for those inputs which match N_{ij}. Then, the sets of guarded assignments GA_i are separately extracted for all sets $S_1, \ldots, S_{m+1}$ of concurrent statements (lines 20-24). The conjunction between the local conditions L of the extracted assignments and additional conditions in form of a certain combination of F_{ij} functions is computed (lines 25-30): For $i < m+1$, the sum G_i of all $F_{ij}, j \in \{1, \ldots, n_i\}$ is used. The ELSE branch ($i = m+1$) is executed, exactly if the selector does not match any of the N_{ij}'s, so that the complemented sum of all G_i is used.

So far, we have only considered extraction for conditional (IF and CASE) statements. For unconditional statements (signal and variable assignments, tristate assignments), such as `<port> <- <expression>`, we simply construct a guarded assignment with a constant $TRUE$ local condition. MIMOLA also permits IF and CASE *expressions* in statements. If `<expression>` is associated with a condition, we use the notion of *guarded expressions*, defined as follows:

3.6.2.1 Definition
A **guarded expression** is a pair (e, F), where e is an RT expression, and F is a local condition.

Transformation of conditions in MIMOLA IF and CASE expressions into local conditions F of guarded expressions is totally analogous to the algorithms in figs. 3.10 and 3.11 and is therefore omitted here.

variable assignment	local condition
r := 0	$P_{c,1} \cdot P_{c,0}$
r := 1	$D_{r>1} \cdot \overline{M_{m,1}} \cdot \overline{M_{m,0}} \cdot (P_{c,1} \cdot P_{c,0})$
r := 2	$\overline{M_{m,1}} \cdot M_{m,0} \cdot \overline{(P_{c,1} \cdot P_{c,0})}$
r := 3	$P_{c,0} \cdot M_{m,1} \cdot \overline{(P_{c,1} \cdot P_{c,0})}$
r := 4	$\overline{P_{c,0}} \cdot M_{m,1} \cdot \overline{(P_{c,1} \cdot P_{c,0})}$

Table 3.3 Variable assignments and local conditions for ComplexModule

3.6.3 Example

Using the above algorithms, we obtain a representation of a behavioral module, which is the set of all permissible guarded port assignments, variable assignments, and tristate assignments. Performing all required Boolean function manipulations by means of BDDs yields a canonical representation of local conditions. We exemplify behavioral analysis for module ComplexModule[3], for which two different MIMOLA descriptions are shown in fig. 3.12. Behavioral analysis yields five guarded assignments to module variable r as shown in table 3.3. Since equivalent Boolean functions have identical BDD representations, the result is independent of the syntax of the module description.

3.7 STRUCTURAL ANALYSIS

After behavioral analysis, the processor *structure* is analyzed, in order to combine *local* guarded assignments to *global* GRTPs across module boundaries. The basic operations in structural analysis are the following:

- For each reference to an IN or INOUT port in the RT expressions of guarded assignments, all possible RT expressions that can be switched to that port are enumerated, while determining the corresponding conditions.

---

[3] This module merely serves the purpose of exposition and would obviously not occur in practice due to its "write-only" behavior.

Instruction-set extraction

Description 1:

```
MODULE ComplexModule(IN c: (1:0));
BEHAVIOR IS
 VAR m: (1:0); -- 2-bit mode register
     r: Byte;  -- data register
CONBEGIN
 IF c.(0) AND c.(1)
  THEN r := 0;
  ELSE CASE m OF
        %00: IF (r > 1) THEN r := 1;
        %01: r := 2;
        ELSE: IF c.(0) THEN r := 3 ELSE r := 4;
       END;
CONEND;
```

Description 2:

```
MODULE ComplexModule(IN c: (1:0));
BEHAVIOR IS
 VAR m: (1:0); -- 2-bit mode register
     r: Byte;  -- data register
CONBEGIN
 CASE c OF
  %00, %01, %10: CONBEGIN
                  IF m.(0) NOR m.(1) THEN
                   IF r > 1 THEN r := 1;
                  CASE m OF
                   %01: r := 2;
                   %1x: r := (IF c.(0) THEN 3 ELSE 4);
                  END;

                 CONEND;
  %11: r := 0;
 END;
CONEND;
```

Figure 3.12 Two descriptions for module ComplexModule

- Each local port variable $P_{p,k}$ in a local condition is substituted by a Boolean function, which represents all alternative control signal combinations that can be switched to bit k of port p.

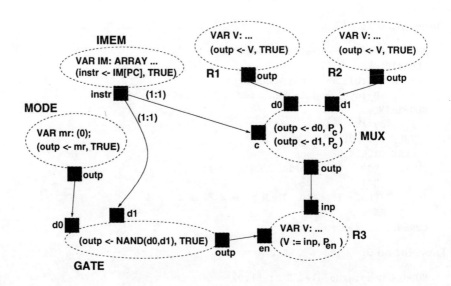

Figure 3.13 Partial processor graph model

Performing these substitutions for each assignment to a module variable or a primary output port yields the set of all valid GRTPs for a target processor. Structural analysis is based on recursive traversal of the graph model presented in section 3.5.

3.7.1 Example

Since structural analysis is relatively complex, we first consider a simple example which illustrates the general idea. Fig. 3.13 shows a partial graph structure with behavioral modules and connections between their ports. Suppose that behavioral analysis has delivered the guarded port and variable assignments shown inside the modules, and we want to extract the GRTPs for variable V of register module R3. We start at the guarded variable assignment[4]

$$(\text{R3.V} := \text{R3.inp}, P_{R3.en})$$

Tracing back the data-flow to port R3.inp through the graph model via module MUX yields the guarded assignments

$$A_1 = (\text{R3.V} := \text{R1.V}, P_{R3.en} \cdot \overline{P_{MUX.c}})$$

---

[4] The notation <name1>.<name2> is used for "global" identification of processor entities. It denotes variable or port <name2> of module <name1>.

Instruction-set extraction

and
$$A_2 = (\text{R3.V} := \text{R2.V}, P_{R3.en} \cdot P_{MUX.c})$$

In order to identify the static conditions for A_1 and A_2, we trace back the data-flow from instruction memory IMEM and mode register MODE to ports R3.en and MUX.c via module GATE, which performs a NAND operation on its inputs:

$$\begin{aligned}
A_1: \quad & P_{R3.en} \cdot \overline{P_{MUX.c}} \\
= \quad & \overline{\overline{M_{MODE.mr} \cdot I_1} \cdot I_1} \\
= \quad & \overline{(\overline{M_{MODE.mr}} \vee \overline{I_1}) \cdot I_1} \\
= \quad & \overline{\overline{M_{MODE.mr}} \cdot I_1} \vee \overline{I_1} \\
= \quad & \overline{I_1} \\
\\
A_2: \quad & P_{R3.en} \cdot P_{MUX.c} \\
= \quad & \overline{\overline{M_{MODE.mr} \cdot I_1} \cdot I_1} \\
= \quad & \overline{(\overline{M_{MODE.mr}} \vee \overline{I_1}) \cdot I_1} \\
= \quad & \overline{M_{MODE.mr}} \cdot I_1
\end{aligned}$$

Thus, we finally obtain the GRTPs

$$(\text{R3.V} := \text{R1.V}, \overline{I_1})$$

and

$$(\text{R3.V} := \text{R2.V}, \overline{M_{MODE.mr}} \cdot I_1)$$

This means, RT R3.V := R1.V is executed, if instruction bit number 1 is set to 0, and R3.V := R2.V is executed, if instruction bit number 1 is 1 and additionally mode register MODE.mr has the value 0.

Systematic construction of GRTPs proceeds in a two-phase *expansion* procedure, which is executed for each guarded variable assignment found during behavioral analysis: In phase one, module port references in RT expressions of assignments are recursively expanded by traversing the processor graph model and enumerating all RT expressions that can be switched to ports by certain control signal settings. Recursion terminates when module variables, primary input ports, or hardwired constants are reached, since these represent the "boundaries" of one machine cycle. In phase two, local conditions of assignments are transformed into RTCs by substituting local port variables. The details are explained in the following two sections.

3.7.2 Expansion of RT expressions

The recursive procedure for expansion of RT expressions is shown in fig. 3.14. For an RT expression E it yields a set GE of guarded expressions. The following cases can occur with respect to the type of E:

E **is a binary expression:** (lines 9-11) The argument expressions E_1, E_2 are expanded recursively, and the resulting sets GE_1, GE_2 of guarded expressions are combined by

$$op(GE_1 \otimes GE_2) := \{(op(e_1, e_2), f_1 \wedge f_2) \mid (e_1, f_1) \in GE_1, (e_2, f_2) \in GE_2\}$$

The logical conjunction of f_1 and f_2 reflects the fact, that two conditions have to be simultaneously satisfied in order to enable a binary operation on expressions e_1 and e_2.

E **is a unary expression:** (lines 13-14) The argument expression E_1 is expanded recursively, and operator op is applied to the resulting set GE_1 of guarded expressions:

$$op(GE_1) := \{(op(e'), f) \mid (e', f) \in GE_1\}$$

E **is a binary constant or a scalar read access to a module variable:** (lines 16, 18) A single guarded expression with a constant $TRUE$ function is generated.

E **is an indexed read access to an array module variable:** The index expression a is recursively expanded, and a read expression is added to GE for each possible index expression.

E **is a read access to a module port:** (line 25) In this case, expansion beyond module boundaries is initiated by calling subroutine EXPANDPORT.

The pseudo-code for EXPANDPORT is given in fig. 3.15, which yields a set of guarded expressions for a module port p. The following cases must be considered:

p **belongs to the interface of a structural module:** (lines 13-14) The module port q is determined, which is connected to p, and expansion is called recursively.

Instruction-set extraction

```
(1)  algorithm EXPANDEXPRESSION
(2)  input: RT expression E
(3)  output: set GE of guarded expressions
(4)  var GE_1, GE_2: set of guarded expressions;
(5)  begin
(6)     GE := ∅;
(7)     case E of
(8)        BINARY EXPRESSION op(E_1, E_2):
(9)           GE_1 := EXPANDEXPRESSION(E_1);
(10)          GE_2 := EXPANDEXPRESSION(E_2);
(11)          GE := op(GE_1 ⊗ GE_2);
(12)       UNARY EXPRESSION op(E_1):
(13)          GE_1 := EXPANDEXPRESSION(E_1);
(14)          GE := op(GE_1);
(15)       BINARY CONSTANT c ∈ {0, 1, x}^+:
(16)          GE := {(c, TRUE)};
(17)       SCALAR READ ACCESS read(r):
(18)          GE := {(read(r), TRUE)};
(19)       INDEXED READ ACCESS read(m, a):
(20)          GE_1 := EXPANDEXPRESSION(a);
(21)          for all (a', f) ∈ GE_1 do
(22)             GE := GE ∪ {(read(m, a'), f)};
(23)          end for
(24)       READ ACCESS TO MODULE PORT p:
(25)          GE := EXPANDPORT(p);
(26)    end case
(27)    return GE;
(28) end algorithm
```

Figure 3.14 Expansion of RT expressions

p belongs to the interface of a behavioral module: (lines 16-22) All RT expressions are expanded, which can be assigned to p by means of a guarded port assignment found during behavioral analysis. For each resulting guarded expression E', the conjunction between its local condition and the local condition of the corresponding port assignment is computed, since both have to be simultaneously true in order to assign E' to p.

```
(1)  algorithm EXPANDPORT
(2)  input: module port p
(3)  output: set GE of guarded expressions
(4)  var q, r: module port;
(5)      Q: set of module ports;
(6)      GE₁: set of guarded expressions;
(7)      T: Boolean function;
(8)      A: set of guarded port assignments;
(9)  begin
(10)     GE := ∅;
(11)     case p of
(12)         STRUCTURAL MODULE PORT:
(13)             q := module port connected to p;
(14)             GE := EXPANDPORT(q);
(15)         BEHAVIORAL MODULE PORT:
(16)             A := all guarded port assignments with destination p;
(17)             for all (p, E, F) ∈ A do
(18)                 GE₁ := EXPANDEXPRESSION(E);
(19)                 for all (E', f) ∈ GE₁ do
(20)                     GE := GE ∪ (E', F ∧ f);
(21)                 end for
(22)             end for
(23)         PRIMARY INPUT PORT:
(24)             GE := {(read(p), TRUE)};
(25)         BUS b:
(26)             Q := set of all ports driving bus b;
(27)             for all q ∈ Q\{p} do
(28)                 GE₁ := EXPANDPORT(q);
(29)                 T := TRUE;
(30)                 for all r ∈ Q\{p, q} do
(31)                     T := T ∧ ADJUSTTRISTATE(r);
(32)                 end for
(33)                 for all (E', f) ∈ GE₁ do
(34)                     GE := GE ∪ {(E', f ∧ T)}
(35)                 end for
(36)             end for
(37)     end case
(38)     return GE;
(39) end algorithm
```

Figure 3.15 Expansion of module ports

Instruction-set extraction 75

p is a primary input port: (line 24) A read access to p with a constant true function is generated.

p is a reference to a tristate bus b: (lines 26-36) This is the most complex case, since expansion must simultaneously take into account all possible bus drivers. Each port $q \neq p$ capable of driving bus b is expanded recursively. In order to permit transmission from q to p without bus conflicts, all remaining bus drivers r must be simultaneously set to a TRISTATE mode. This is accomplished by means of subroutine ADJUSTTRISTATE which is defined as follows:

$$\text{AdjustTristate}(r) = \bigvee_{F \in \mathcal{F}_{tri}} F$$

where $\mathcal{F}_{tri}$ is the set of all Boolean functions F, so that $(r, TRISTATE, F)$ is a tristate assignment to port r.

3.7.3 Expansion of conditions

Local conditions derived during behavioral analysis refer to local port variables, local mode register variables, and local dynamic conditions. In order to identify *global* static and dynamic conditions (RTCs) and to reveal possible intra-RT conflicts, local conditions need to be expanded across module boundaries, similarly to RT expressions. After expansion, conditions only refer to instruction bit variables, (global) mode register variables, and (global) dynamic conditions. The algorithm in fig. 3.16 takes a local condition F and iteratively generates a (global) RTC by performing appropriate substitutions of each variable of F. For constant F, no expansion is necessary. Otherwise, the *support set* of F is determined (line 7), i.e. the set of all Boolean variables x_i, on which F actually depends:

$$\text{Support}(F : \{0,1\}^n \to \{0,1\}) =$$
$$\{x_i \mid \exists \ z \in \{0,1\}^n : \ F|_{x_i=1}(z) \neq F|_{x_i=0}(z)\}$$

Then, all Boolean variables in the support set are consecutively considered. For each $y \in \text{Support}(F)$, a Boolean function G is constructed, by which y must be substituted in order to obtain a global condition. Function G represents conditions generated by other modules, which can be switched to the "local wire" represented by variable y. Three cases are distinguished:

(1) y is a port variable: (lines 11-29) Let $y = P_{p,k}$, i.e. y refers to bit k of module port p. First, all possible replacements of p are determined by

(1) **algorithm** EXPANDCONDITION
(2) **input:** Boolean function F;
(3) **output:** Boolean function F';
(4) **var** Y: **set of** Boolean variables;
(5) $\quad G, H$: Boolean function; GE: **set of** guarded expressions;
(6) **begin**
(7) $\quad F' := F; Y := \text{SUPPORT}(F')$;
(8) $\quad$ **for all** $y \in Y$ **do**
(9) $\quad\quad G := FALSE$;
(10) $\quad\quad$ **case** y **of**
(11) $\quad\quad\quad$ LOCAL PORT VARIABLE $P_{p,k}$:
(12) $\quad\quad\quad\quad GE := \text{EXPANDPORT}(p)$;
(13) $\quad\quad\quad\quad$ **for all** $(e, f) \in GE$ **do**
(14) $\quad\quad\quad\quad\quad$ **case** $e.(k)$ **of**
(15) $\quad\quad\quad\quad\quad\quad$ BIT j OF INSTRUCTION MEMORY:
(16) $\quad\quad\quad\quad\quad\quad\quad H := I_j \;\wedge\; \text{EXPANDCONDITION}(f)$;
(17) $\quad\quad\quad\quad\quad\quad$ BIT j OF MODULE VARIABLE v:
(18) $\quad\quad\quad\quad\quad\quad\quad H := M_{v,j} \;\wedge\; \text{EXPANDCONDITION}(f)$;
(19) $\quad\quad\quad\quad\quad\quad$ COMPARISON EXPRESSION $c = op(e_1, e_2)$:
(20) $\quad\quad\quad\quad\quad\quad\quad H := D_c \;\wedge\; \text{EXPANDCONDITION}(f)$;
(21) $\quad\quad\quad\quad\quad\quad$ CONSTANT BIT $b \in \{0, 1, x\}$:
(22) $\quad\quad\quad\quad\quad\quad\quad$ **case** b **of**
(23) $\quad\quad\quad\quad\quad\quad\quad\quad$ '1': $H := \text{EXPANDCONDITION}(f)$;
(24) $\quad\quad\quad\quad\quad\quad\quad\quad$ '0': $H := FALSE$;
(25) $\quad\quad\quad\quad\quad\quad\quad\quad$ 'x': $H := \text{EXPANDCONDITION}(f) \wedge \mathcal{X}$;
(26) $\quad\quad\quad\quad\quad\quad\quad$ **end case**
(27) $\quad\quad\quad\quad\quad$ **end case**
(28) $\quad\quad\quad\quad\quad G := G \vee H$;
(29) $\quad\quad\quad\quad$ **end for**
(30) $\quad\quad\quad$ LOCAL MODE REGISTER VARIABLE $M_{w_j,k}$: $G := M_{w_j,k}$;
(31) $\quad\quad\quad$ LOCAL COMPARISON VARIABLE D_c:
(32) $\quad\quad\quad\quad GE := \text{EXPANDEXPRESSION}(c)$;
(33) $\quad\quad\quad\quad$ **for all** $(c', f) \in GE$ **do**
(34) $\quad\quad\quad\quad\quad G := G \vee (D_{c'} \;\wedge\; \text{EXPANDCONDITION}(f))$;
(35) $\quad\quad\quad\quad$ **end for**
(36) $\quad\quad$ **end case**
(37) $\quad\quad F' := F'|_{y=G}$;
(38) $\quad$ **end for**
(39) $\quad$ **return** F';
(40) **end algorithm**

Figure 3.16 Expansion of conditions

means of subroutine EXPANDPORT (see fig. 3.15). The result is a set GE of guarded expressions (e, f). Each e represents an alternative substitution for y, presuming f evaluates to $TRUE$. Let $e.(k)$ denote the k-th bit of expression e, i.e. the "bit line" of e connected to bit k of port p. A substitution function H for $e.(k)$ is generated, and all alternative substitutions are "collected" in function G (line 28). The following situations can occur with respect to $e.(k)$:

1. $e.(k)$ refers to bit j of the (labelled) instruction memory (line 15): Then, y can be substituted by instruction bit variable I_j, exactly if the guard f of e evaluates to $TRUE$. Therefore, f is recursively expanded, and the conjunction with I_j is computed. The cases that $e.(k)$ refers to a mode register or to a dynamic condition are treated analogously (lines 17, 19).

2. $e.(k)$ is a binary constant b (lines 21-26): In this situation, the gap between the two-valued logic of Boolean functions and the three-valued logic $\{0, 1, x\}$ of MIMOLA must be bridged. The case $b = x$ occurs, if port p is connected to a module, whose output behavior is incompletely specified (see e.g. the decoder in fig. 2.6), so that the value of variable y becomes undefined for certain machine states. This problem can be circumvented by introducing a distinguished Boolean variable $\mathcal{X}$, which represents an "undefined condition". Introduction of $\mathcal{X}$ permits to perform expansion using two-valued logic only. Those RTCs, which have $\mathcal{X}$ in their support set after expansion is finished, can be eliminated later as described in section 3.8.2. We define, that y evaluates to $TRUE$ for $b = x$, exactly if $\mathcal{X} = 1$ and the guard f evaluates to $TRUE$. For $b = 1$, only f must be $TRUE$. Therefore, the substitution H in case $b = 1$ is the expansion of f, while for $b = x$, the expansion of f is combined with $\mathcal{X}$. For $b = 0$, variable y can never evaluate to $TRUE$, so that only a "neutral element" for the disjunction in line 28 needs to be generated.

(2) y is a mode register variable: (line 30) No further expansion is necessary, and the variable is kept.

(3) y is a comparison variable: (lines 31-35) Let $y = D_c$. RT expression c is expanded by subroutine EXPANDEXPRESSION (see fig. 3.14), resulting in a set of guarded expressions (c', f). As above, the conditions f are recursively expanded, and the disjunction of all alternatives is computed.

Finally, function G represents all alternative substitutions for y. The substitution is executed (line 37), and the next variable y is processed.

Instruction-set extraction for a complete processor model is accomplished by calling the above RT expression and RT condition expansion algorithms for each guarded assignment found during behavioral analysis. The result is the set of all GRTPs In case of a non-satisfiable register transfer conditions an intra-RT conflict is exposed, and the corresponding GRTPs are deleted from the GRTP base.

3.8 POSTPROCESSING

3.8.1 Computation of no-operations

As explained in section 3.4.4, the instruction-set model must also account for no-operations. Once the set of all GRTPs is known, NOPs for module variables are easily obtained by the algorithm in fig. 3.17. For a module variable $\hat{v}$, algorithm COMPUTENOP determines a no-operation by computing the condition g, under which v is *not* written by any GRTP. In case of multiport memories, a NOP can be determined for each memory write port, which is omitted here for sake of brevity. Special cases of NOP conditions are $g \equiv 0$, i.e. variable v is mandatorily written in each machine cycle, and $g \equiv 1$, i.e. v cannot be written at all and can be considered as a read-only component.

(1) **algorithm** COMPUTENOP
(2) **input:** module variable v
(3) **output:** NOP for v
(4) **var** f: Boolean function;
(5) R: set of GRTPs;
(6) **begin**
(7) $f := FALSE$;
(8) $R :=$ all GRTPs with destination v;
(9)   **for all** $(\mathcal{R}, F) \in R$ **do** $f := f \vee F$;
(10)   **return** $(v, \overline{f})$;
(11) **end algorithm**

Figure 3.17 Computation of NOPs

3.8.2 Decomposition of RT conditions

In general, register transfer conditions of GRTPs and NOPs represent a number of *alternatives* with respect to control signal settings. In section 3.4.3 we have exemplified how such alternatives can arise for instance due to alternative routes in a processor data-path. During code generation, exactly one of the possible alternatives is selected for each generated RT, typically so as to maximize parallelism in a machine program. In order to enable this selection and to identify the corresponding partial instructions, mode register states, and dynamic conditions, it is necessary to *explicitly* enumerate all alternative control signal settings for each RTC. This can be accomplished based on a BDD representation of an RTC.

From the BDD for a Boolean function (an RTC) $F : \{0,1\}^n \rightarrow \{0,1\}$, one can derive a sum-of-products representation of F as follows: All different paths $S = (s_1, \ldots, s_k, 1)$ from the BDD root node to the leaf node labelled as "1" are enumerated. For each path S, a product term $P = p_1 \cdot \ldots \cdot p_k$ is generated, such that $p_i = \overline{s_i}$, if s_i was reached via a zero edge, and $p_i = s_i$, otherwise. Obviously, F can be written as

$$F = P_1 \vee \ldots \vee P_m$$

where each P_i corresponds to one path. For all $z \in \{0,1\}^n$ and $i \in \{1, \ldots, m\}$, $P_i(z) = 1$ implies $F(z) = 1$, so that the P_i's are *implicants* of F. Thus, each product term P_i can be regarded as an alternative for fulfilling the RTC F: Ignoring don't cares, P_i evaluates to $TRUE$ for exactly one setting of control signals.

A special case is, that P_i comprises an $\mathcal{X}$ or $\overline{\mathcal{X}}$ literal. We had introduced variable $\mathcal{X}$ in order to cope with possible undefined conditions resulting from incompletely defined MIMOLA modules (section 3.7.3, fig. 3.16). If P_i comprises $\mathcal{X}$ or $\overline{\mathcal{X}}$, we must consider P_i as an "illegal" alternative, since the actual value of $\mathcal{X}$ is unpredictable. Such illegal alternatives are discarded by deleting P_i from the sum-of-products. By decomposing each (legal) product term into a representation

$$P_i = I \cdot M \cdot D$$

where I is a product term on instruction bit variables, M is a product term on mode register variables, and D is a product term on dynamic condition variables, the necessary partial instructions, mode register states, and dynamic conditions are made explicit. In turn, this provides the information required for code generation: In order to select a certain GRTP for a certain control step, the code generator must emit the partial instruction corresponding to

I. If $M \neq 1$, then the code generator must ensure, that the mode register states encoded in M are already present, i.e. code must be generated which appropriately loads mode registers in an earlier control step. If $D \neq 1$, then the GRTP can be used for code selection for source code statements involving run-time decisions, e.g. IF-statements.

A BDD for a function F is generally no *minimal* representation for F. In particular, the product terms derived as described above are not necessarily *prime implicants* of F, and thus might prescribe certain control signal settings, which are actually don't care. As an example, suppose that the following sum-of-products has been derived from a BDD:

$$F(I_2, I_1, I_0) = \overline{I_2} \cdot \overline{I_0} \vee I_1 \cdot \overline{I_0} \vee I_1 \cdot I_0$$

F evaluates to $TRUE$ for the partial instructions 0x0, x10, and x11. Alternatively, F can be written as

$$F(I_2, I_1, I_0) = \overline{I_2} \cdot \overline{I_0} \vee I_1$$

which evaluates to $TRUE$ for 0x0 and x1x. In order to obtain such a more succinct representation it is favorable to compute a sum-of-products representation of F, which consists of all prime implicants of F. Since prime implicants are the shortest possible implicants of a function, it is guaranteed that all don't cares are removed. Computation of prime implicants is mandatory in presence of $\mathcal{X}$ literals: For instance, if variable I_0 in the above example function were substituted by $\mathcal{X}$, then all three alternatives would be regarded as illegal when using the first sum-of-products representation. The second representation with prime implicants, however, reveals that product term "I_1" is actually legal, and therefore must not be discarded.

3.9 EXPERIMENTAL RESULTS

Instruction-set extraction, as introduced in the previous sections, is not dedicated towards DSP processors only, but can be basically applied to any MIMOLA processor model. In order to demonstrate feasibility of ISE in the context of retargetable code generation, we have applied ISE to a number of different target architectures, using the BDD package from [BRB90]. "demo", "simplecpu", "jmpcpu" are simple benchmark structures taken from [Biek95], "ref" is a processor described in [BBH+94], which was originally designed for prime factor decomposition. The processors "manocpu" and "tanenbaum" are education-purpose processors described in [Mano93] and [Tane90], respectively.

Instruction-set extraction

GRTP	partial instr. (bits 20..0) 21111111111 098765432109876543210
PC.R := INCR PC.R	xxxxxxxxxxxxxxxxxxxxx
REG.R := inp	xxxxxxxxxxxxxxxxx011x
REG.R := IM.storage[PC.R].(20:13)	xxxxxxxxxxxxxxxxx001x
REG.R := RAM.storage[IM.storage[PC.R].(12:5)]	xxxxxxxxxxxxxxxxx1x1x
REG.R := REG.R - inp	xxxxxxxxxxxxxxxxx0101
REG.R := REG.R - IM.storage[PC.R].(20:13)	xxxxxxxxxxxxxxxxx0001
REG.R := REG.R - RAM.storage[IM.storage[PC.R].(12:5)]	xxxxxxxxxxxxxxxxx1x01
REG.R := REG.R + inp	xxxxxxxxxxxxxxxxx0100
REG.R := REG.R + IM.storage[PC.R].(20:13)	xxxxxxxxxxxxxxxxx0000
REG.R := REG.R + RAM.storage[IM.storage[PC.R].(12:5)]	xxxxxxxxxxxxxxxxx1x00
RAM.storage[IM.storage[PC.R].(12:5)] := REG.R	xxxxxxxxxxxxxxxx1xxxx
outp ← REG.R	xxxxxxxxxxxxxxxxxxxxx
NOP	
NOP IM.storage	xxxxxxxxxxxxxxxxxxxxx
NOP RAM.storage	xxxxxxxxxxxxxxxx0xxxx

Table 3.4 Extracted operations for SimpleProcessor

"simpleprocessor" is the example processor from section 2.1, for which all extracted GRTPs and NOPs are shown in table 3.4. "TMS320C25" and "DBB" are the architectures from the application studies in chapter 2. The two latter architectures have been modelled in behavioral and mixed-level style, respectively.

The overall results are listed in table 3.5. The complexity of the different targets is outlined by columns 2 (lines of code) and 3 (instruction word-length). Columns 4 and 5 give information about BDD usage, i.e. the number of different BDD variables and total number of BDD nodes required for the complete instruction-set model. The numbers of GRTPs and NOPs (decomposed, prime implicants computed) are shown in columns 6 and 7. Finally, column 8 gives the required CPU time on a SPARC-20 workstation. As indicated by the experimental results, the CPU time required for ISE essentially depends on the amount of BDD usage. In turn, this depends on certain factors, which are not directly related to the model size. These include the **degree of combinational chaining** in the data-path and the **degree of instruction encoding**.

architecture	LOC	i-length	BDD vars	BDD nodes	GRTPs	NOPs	CPU sec
demo	257	84	36	40,832	124	5	346
simplecpu	170	48	17	640	12	2	0.2
jmpcpu	291	55	17	22	248	3	0.4
ref	215	55	28	28,797	880	4	25
manocpu	513	50	55	746	123	14	1.4
tanenbaum	356	32	25	640	90	5	1.4
simpleprocessor	61	21	11	97	12	2	0.2
TMS320C25	1331	16	37	48,169	510	2051	134
DBB	169	41	37	577	38	38	0.6

Table 3.5 Experimental results for instruction-set extraction

Nevertheless, the CPU time for ISE can be regarded as acceptable in all cases, because ISE is a frontend procedure, which has to be performed only once for each particular target architecture. The instruction-set model created by ISE forms the basis for the code generation techniques presented in the following two chapters.

3.10 ISE AS A VALIDATION PROCEDURE

The comprehensive analysis of control signals in ISE by means of Boolean functions also enables some further applications of ISE with respect to validation of processor models, which turns out to be particularly useful during development of new processor models. We briefly mention the following:

Behavioral module descriptions are obviously only correct, if concurrent assignments of different RT expressions to the same destination are **mutually exclusive**. Mutual exclusion of two assignments with RTCs F_1, F_2 can be checked by computing

$$G = (F_1 \Rightarrow \overline{F_2}) \land (F_2 \Rightarrow \overline{F_1}) = \overline{F_1 \land F_2}$$

$G \equiv 1$ is a sufficient condition for mutual exclusion of the assignments.

Instruction-set extraction 83

One can check whether or not the output behavior of a module is **completely defined**, i.e. whether or not there exists a (local) control signal setting for which the value of an output port p is undefined. The output behavior is completely defined, exactly if the disjunction of all RTCs of assignments to p is a constant $TRUE$ function.

ISE can be used to verify that a certain hardware structure meets a given instruction-set specification. Since ISE enumerates all possible RT patterns, also **hidden instructions**, i.e. instructions not intended by the processor designer, are detected.

4

CODE GENERATION

This chapter is concerned with mapping of algorithms to processor-specific RT patterns, which have been obtained by ISE. In this way, code generation algorithms can abstract from unnecessary hardware details. In turn, this enables the use of advanced and efficient code generation techniques. The main problem, however, is to cope with inhomogeneous architectures of DSPs, while achieving high code quality and retaining retargetability. We discuss how this problem is treated in related work, and we propose a partially integrated approach based on a fine-grained definition of code generation phases. This approach employs tree parsing as the key technique for a complete and retargetable code generation procedure for DSPs.

4.1 TARGET ARCHITECTURE STYLES

The efficacy of a certain code generation methodology, including ordering and coupling of compilation phases (code selection, register allocation, scheduling), cannot be assessed without considering the characteristics of the underlying *target architecture*. Most of the work done in compiler construction is based on the assumption of a *homogeneous architecture*. An architecture is called **homogeneous**, if there is nearly unrestricted parallelism between functional units, and all registers are interchangeable.

For machines with interchangeable registers and a low degree of parallelism, register allocation is frequently mapped to a graph coloring problem. Variables may share a register, if their *lifetimes* (the interval between definition and use) do not overlap. The lifetime constraints can be represented by an *in-*

terference graph, which contains a node for each variable and an edge for each lifetime overlap. Spilling can be avoided, exactly if the interference graph is k-colorable, where k is the number of registers. In special cases, k-colorability of the interference graph can be decided in linear time, for instance, if exactly one assignment is present for each variable ("single assignment property"). In this case the interference graph is an *interval graph* [Golu80], for which the *left-edge algorithm* [KuPa87] achieves optimal register allocation. In general, graph colorability is an NP-complete problem [GaJo79], so that different heuristics have been proposed [Chai82, Brig92].

In presence of instruction-level parallelism, the interdependence between register allocation and scheduling is much stronger. More advanced approaches [GoHs88, BEH91, BSBC95] therefore integrate both phases by including early estimations of effects of register allocation on scheduling and vice versa. Register allocation also interacts with code selection: Depending on resource availability, it can be favorable to consider *rematerialization* of values [BCT92]. Finally, trade-offs can be made among code selection and scheduling: *Incremental tree height reduction* [NiPo91] re-structures expressions on-the-fly by application of algebraic rules, so as to optimize utilization of parallel functional units. Mutation Scheduling [NND95] aims at integrating all three phases into a unified framework.

Unfortunately, the above techniques do not provide sufficient support for DSPs, which tend to show an *inhomogeneous architecture*. An architecture is called **inhomogeneous**, if it comprises special-purpose registers, dedicated to inputs and outputs of specific functional units (FUs). Further characteristics of inhomogeneous architectures are *restricted parallelism* between functional units, due to instruction encoding, and FU chaining. For such architectures, binding of operations to FUs implies assigning values to certain registers and vice versa. Thus, register allocation for inhomogeneous architectures should not be dominated by lifetime analysis, but by the costs of moving data between special-purpose registers and FUs. Moreover, the spill costs for special-purpose registers in general depend on their position in the data-path.

4.2 PROGRAM REPRESENTATIONS

Besides the characteristics of the target architecture, also the selected *intermediate program representation* and the *scope of optimization* strongly influence

Code generation

code generation techniques. First of all, *local* and *global* techniques can be distinguished, based on the notion of "basic blocks":

4.2.1 Definition
A **basic block** is a sequence of assignments with at most one conditional assignment. If a conditional assignment is present, then it appears at the last position in the sequence.

If basic blocks are considered at the *source level*, assignments refer to program variables and source language operators. Alternatively, basic blocks can be defined at the *RT level*, where assignments refer to physical memory locations and correspond to available RT patterns. In local techniques, code optimization does not cross basic block boundaries, while global techniques aim at higher utilization of parallelism through "moving" assignments between basic blocks, dependent on resource availability. However, global techniques may suffer from a large increase in code size. In our approach, we therefore focus on local code generation. Within the scope of basic blocks, we can further distinguish between *graph-based* and *tree-based* techniques. A basic block having the single assignment property can be represented by a *data-flow graph*:

4.2.2 Definition
The **data-flow graph** (DFG) for a basic block BB is a directed acyclic graph $G = (V, E)$. Each node $v \in V$ represents either a primary input to BB or an assignment in BB. Each edge $e = (v_1, v_2) \in E$ represents data-flow in BB, i.e. the assignment represented by v_2 consumes data produced by the assignment (the primary input) represented by v_1. A data-flow graph $G = (V, E)$ is a **data-flow tree** (DFT), if $|E| = |V| - 1$ and all nodes $v \in V$ have at most one outgoing edge.

A basic block and its DFG are depicted in fig. 4.1 a) and b). DFG nodes with more than one outgoing edge (i.e. values multiply referenced in a basic block) are called *common subexpressions*. The problem of optimal code generation for DFGs with common subexpressions is NP-hard, even for idealized machines [BrSe76, AJU77]. Therefore, the most widespread approach to cope with common subexpressions is to decompose DFGs into forests of DFTs [ASU86], as shown in fig. 4.1 c). A separate DFT is created for each common subexpression, and code is generated, which evaluates the DFT into a certain register or storage location. All references to a common subexpression are replaced

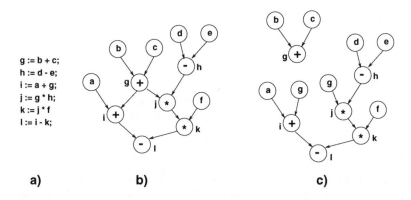

Figure 4.1 a) basic block, b) corresponding DFG, c) corresponding forest of DFTs

by *separate* references to the location of the corresponding DFT. The penalty of DFG-to-DFT decomposition with respect to code quality is, that optimality can no longer be guaranteed even if optimal code for DFTs is generated, because communication of values between DFTs may cause superfluous move instructions. On the other hand, DFTs are much easier to handle. Code generation for DFTs is mostly considered as a problem of *tree covering*, during which a minimal number of instances of available tree patterns (instruction patterns) are selected. In contrast to general DFGs, optimal DFT covers can be computed in linear time by tree matching with dynamic programming [AGT89]. Here, we follow the tree covering approach as well, because of the following reasons:

1. As we point out in the following section, those approaches which perform code generation on general DFGs suffer from **too high computational complexity** and are restricted to a small range of architectures.

2. It is necessary to include **transformation rules** into retargetable code generation. In general, a number of alternative code transformations have to be tried in order to cope with peculiarities of the target machine. This is, however, only feasible, if code generation for each of the alternatives can be done efficiently.

3. Standard **code generator generators** can be employed, which guarantee a high degree of automation in retargeting, as well as short turnaround times.

Before presenting our approach in more detail, we investigate how code generation for inhomogeneous architectures is solved in related work.

4.3 RELATED WORK

Greedy approach

Code selection and allocation of special-purpose registers in the MSSQ compiler (cf. section 2.2). are implemented by *recursive descent* (RD) parsing of DFTs with respect to a "grammar" represented by the connection operation graph (COG). RD parsing constructs a *leftmost derivation* (see [AhUl72] for formal details) of a given input string by means of backtracking. Since RD parsing operates on *linear* inputs (i.e. strings), DFTs need to be sequentialized. The sequences also immediately induce the schedule of generated RTs. This type of code generation can be considered as *greedy*, because the first successful derivation is always accepted.

Targeted approach

In the targeted approach, "expert knowledge" about the target processor is exploited to obtain high-quality code. On the other hand, incorporation of expert knowledge either significantly restricts the class of possible target processors, or demands for a considerable amount of manual interaction. A branch-and-bound scheduling technique for *accumulator-based machines* has been treated by Liao [LDK+95b]. Although Liao's technique is likely to produce good results for a restricted machine class, it assumes that the input DFG already consists of machine instructions, i.e. code selection and register allocation have been performed by some preprocessor. This problem is avoided in Liao's alternative approach [LDK+95c], in which code generation for DFGs and accumulator-based machines is transformed into a formal optimization problem. The problem formulation covers both code selection and scheduling, but so far no practical results could be produced, presumably due to too high runtime requirements. Code generation in the CodeSyn compiler [LMP94a, LMP94b] follows a strict separation of code selection and register allocation phases: Dynamic programming is used to compute DFT covers by elements of a user-specified database of RT patterns. Then, register allocation takes place. It is assumed that the registers, from/to which selected RTs read and write data, are not yet completely fixed, but that each of these registers can be selected from a certain set of *candidates*. The objective of register allocation is to provide efficient data transport between scheduled RTs. The CodeSyn approach is effective, whenever sufficiently large candidate register sets are actually present, which implies a large optimization potential for register allocation. Otherwise, the unfavorable effects of phase separation will become dominant. A special case of the targeted approach is the methodology presented in [LPCJ95], which uses a macro-expansion procedure in order to generate target processor code. Expansion is guided by manually specified translation rules. The major drawback of the rule-based approach is its lack of automation.

Data routing approach

Similar to CodeSyn, the data routing approach, as exemplified in the CBC and CHESS compilers [FaKn93, LVKS+95], separates code selection from register allocation. In CBC, code selection is performed by a tree parser, which is automatically generated from a processor model in the nML language [FVM95]. Transportation of data between selected RT patterns ("data routing") and scheduling are performed by Hartmann's technique [Hart92], which presumably is too inefficient for complex architectures. In CHESS, code selection is performed by heuristic assignment of source code operations to available functional units, while giving priority to chained operations. Encoding restrictions are taken into account by means of an *instruction set graph*. After code selection, data routing takes place. In contrast to other approaches, data routing in CHESS explores a global search space. However, it restricts CHESS to *load-store architectures*. The scheduling freedom left after code selection and data routing is exploited during a final *list scheduling* phase. The advantage of CHESS is easy retargetability within the scope of load-store architectures. Whether or not the powerful data routing techniques in CHESS compensate the disadvantages of suboptimal code selection and phase separation is yet to be demonstrated.

Integrated approach

Several researchers have investigated code generation methods, which *simultaneously* solve different code generation subtasks. Wess' approach was the first, that integrated code selection and register allocation for inhomogeneous data-paths [Wess92], by taking into account data transport costs already during covering of DFTs. This is accomplished by means of *trellis diagrams*. A trellis diagram is constructed for each arithmetic operation available in the target processor, and it encodes the sets of possible input/output registers for the operation (similar to CodeSyn's candidate register sets). Additional trellis diagrams are designated to pure data moves between registers. Code generation for DFTs is performed by a dynamic programming algorithm. The advantage of Wess' approach is that it generates provably optimal DFT covers in presence

Code generation

of special-purpose registers. On the other hand, no mechanism is provided for automatic construction of the underlying trellis diagrams from more common processor models. Araujo's approach [ArMa95], assumes that an instruction-set specification in form of a *tree grammar* is given. From the grammar, a processor-specific tree parser is generated by means of a code generator generator. Since tree grammars may include *register-specific patterns*, move operations are minimized simultaneously with the number of instructions required to cover DFT operations. It is shown, that for a certain class of architectures spill-free schedules can always be derived from covered DFTs. Code generation for more general architectures is, however, not considered. The highest degree of phase coupling is achieved in the *Integer Programming* (IP) formulation by Wilson et al. [WGHB94]. Both DFG representations of basic blocks and the target architecture are encoded into a set of linear constraints. The resulting Integer Program is solved by standard tools, where the objective is to minimize the number of control steps. Although the resulting code is guaranteed to be optimal, but retargeting requires a large amount of manual work. Moreover, due to the high computation time for IP solving, only very simple targets have been handled so far. The same also holds for other recent formal approaches to optimal code generation for DFGs [LaCe93, Mahm96, MaTe96].

4.4 THE CODE GENERATION PROCEDURE

The organization of code generation, as proposed in this book, is driven by the general demand to compile code for any target that fits the architecture class defined in section 1.5.3. Furthermore, short turnaround times for retargeting must be guaranteed, and high attention must be paid to code optimization. In order to meet these demands, we follow a partially integrated approach, based on the following fine-grained definition of code generation phases.

4.4.1 Definition of code generation phases

Code selection phases

- **Tree selection:** For each DFT in the intermediate program representation, a set of alternative, semantically equivalent DFTs is constructed by applying transformation rules. The purpose of these transformations is to cope with peculiarities in the target machine and to exploit algebraic rules in order to improve code quality. Tree selection determines the cheapest alternative, based on an estimation of the number of required machine instructions.

- **RTP selection:** For a given DFT, RTP selection determines an optimal implementation, which consists of RT patterns available on the target processor.

- **Version selection:** If alternative partial instructions exist for selected RTPs, version selection determines that partial instruction for each RTP, which locally maximizes instruction-level parallelism.

Register allocation phases

- **Register binding:** Register binding assigns program values and intermediate results in DFTs to specific registers or register files, while minimizing the amount of data moves between functional units.

- **Address assignment:** For values bound to register files, that are equipped with a dedicated address generation unit, a mapping to physical addresses is computed. The goal is to maximize utilization of parallel address generation hardware, that is, minimization of additional address generation code.

Scheduling phases

- **Source-level scheduling:** The order, in which DFTs in the intermediate representation are processed, must preserve semantical correctness of the program. Source-level scheduling determines such an order, which obeys control and data dependencies between DFTs.

- **RT scheduling:** DFTs covered by RTPs must be transformed into vertical code, i.e. RTL basic blocks, in order to determine the spill requirements. RT scheduling aims at minimizing spill costs and inserts possibly necessary spill and reload code.

- **Compaction:** Compaction is the task of assigning RTs in an RTL basic block to control steps. The goal is to maximize instruction-level parallelism, while obeying inter-RT conflicts and dependencies.

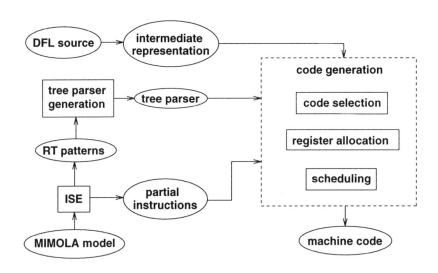

Figure 4.2 Overview of the retargetable compilation process

4.4.2 Organization of code generation

Fig. 4.2 gives an overview of the compilation process, including preprocessing steps: The DFL source program is translated into a set of partially control and data-dependent DFTs, which are passed to the code generator. The MIMOLA model of the target processor is analyzed by ISE, and a processor-specific tree parser is generated from the extracted RT patterns. The tree parser is used in code selection, register allocation, and scheduling. During the code compaction phase of scheduling, also the partial instructions extracted for each RT are taken into account.

The organization of code generation is as follows (fig. 4.3):

1. Intermediate code generation: The DFL source program is translated into a set of partially interdependent DFTs. DFTs capture data dependencies between primitive source code operations. These data dependencies are either implicit in the source code assignments, or are derived by means of *data-flow analysis* between assignments. Common subexpressions are resolved by DFG-to-DFT decomposition. Intermediate code generation also implies partial register binding. We distinguish *foreground* and *background* registers. Foreground registers are either single registers or register files with limited capacity. Background registers are on-chip memories of unlimited[1] capacity. Certain program values, including variables and arrays, are a priori bound to background registers, which serve as "home locations" for these values. In presence of multiple background register files, this binding can be guided by the user. Also common subexpressions are bound to background registers. After intermediate code generation, the primitive source code entities are DFTs of maximum size, which "communicate" via background registers.

2. Source-level scheduling: A sequential ordering of generated DFTs is computed, which preserves semantical correctness by obeying control and data dependencies between DFTs. In case of control dependencies, the ordering is determined by the branching hardware of the target machine. Data-dependent DFTs are sorted topologically. According to the computed ordering, the DFTs are step-by-step passed to the subsequent code generation phases.

3. DFT transformation and selection: In order to capture special features of the target machine and to improve code quality by exploitation of algebraic rules, several semantically equivalent alternatives for each DFT in the intermediate representation are generated. The relative quality of different alternatives is estimated by the number of RTs required to implement a DFT, which is known after RT scheduling. After all alternatives have been tried, the cheapest tree is selected.

4. RTP selection and register binding: DFTs are covered by available RT patterns. This is accomplished by means of an automatically generated tree parser. The result is a *register transfer tree*, in which nodes represent RT patterns, and edges represent data dependencies.

5. RT scheduling: RT trees are translated into vertical code by determining an appropriate tree evaluation order, which minimizes spill code. If spilling cannot be avoided, a spill candidate is heuristically determined, and the vertical code is augmented by spill and reload code. If necessary, additional code is generated which ensures the correct mode register states.

6. Address assignment: A detailed mapping of values assigned to background registers is computed, and the vertical code is augmented by the necessary RTs for address generation.

---

[1] In practice, this means that background registers are assumed to be sufficiently large, so that spilling is never necessary. The distinction between foreground and background registers is made by passing the minimum background memory size as a parameter to the code generator. This is feasible, because in realistic architectures foreground and background registers can be distinguished very clearly.

Code generation

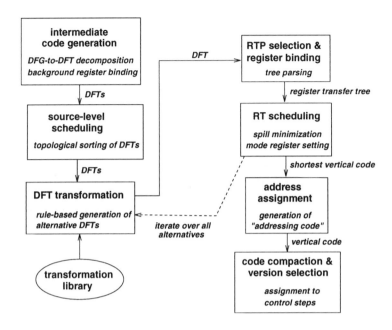

Figure 4.3 Organization of code generation

7. **Code compaction and version selection:** After address assignment, the complete RTL basic blocks for implementing DFTs are known. While obeying inter-RT dependencies and conflicts, RTs are assigned to control steps. Alternative partial instructions are taken into account in order to maximize parallelism and to avoid undesired side effects¿ The result is executable machine code.

The remainder of this chapter provides a detailed explanation of code generation down to RT scheduling. Address assignment and code compaction are described in chapter 5.

4.5 DFL LANGUAGE ELEMENTS

As already motivated in the introductory chapter, we use the data-flow language DFL for description of DSP algorithms. This choice has been made, because DFL currently is the only high-level programming language with built-in support for DSP applications. Here, we outline the specific features of DFL, as far as code generation techniques are concerned. The exact language specification can be found in the manual [DFL93].

4.5.1 Bit-true data types

Unlike in standard programming languages, having *implementation-dependent* data types, type specifications in DFL are annotated with bit widths. For instance, "`signal s : int<12>;`" declares signal s as a 12-bit wide integer number in two's complement representation. As a generalization of integer numbers, DFL also offers *fixed-point* numbers as a built-in data type. Declarations of fixed-point numbers specify a word-length and a *fractional length*, i.e. the number of bits right to the binary point. DFL also comprises unsigned integer and fixed-point numbers as well as a Boolean type.

4.5.2 DSP-specific operators

For general-purpose processors, high-level language programs are mostly written in such a way that arithmetic overflows are excluded. In the area of fixed-point DSP, however, overflows may occur, but they must not cause delays in the computation. Therefore, DFL permits specification of *overflow and quantization characteristics* of arithmetic operators. Often, overflows are treated by *saturation*, i.e. results causing positive (negative) overflow are replaced by the largest (smallest) representable number. In case of fixed-point numbers, also the *rounding* behavior of operators needs to be defined. The consequences of finite word-length effects on code generation are twofold: Firstly, code selection must ensure, that RTs are selected in accordance with the specified characteristics. This can be realized rather simply by annotating operators both in hardware and software with saturation and rounding modes. Secondly, code must be generated, which sets appropriate modes in the target processor, that is, mode registers need to be loaded appropriately.

DFL programs are supposed to run in an *infinite loop*, i.e. the specified computation is repeated for each set of new inputs, which arrive under a certain *sampling rate*. In chapter 1, we have already mentioned the FIR filter example, which realizes the equation

$$y(n) = a0 \cdot x(n) + a1 \cdot x(n-1) + a2 \cdot x(n-2)$$

In each sample period n, the "delayed" input values $x(n-1)$ and $x(n-2)$ from sample periods $n-1$ and $n-2$ are referenced. Thus, code must be generated, which ensures that each delayed value can be accessed by the same instructions in each period, i.e. the delay line is properly updated. In DFL, the FIR filter behavior is expressed by

```
y = a0 * x + a1 * x@1 + a2 * x@2;
```

The delay operator "@i" accesses the i-th previous value of a signal, and implicitly describes the necessary update operation.

4.5.3 Programs

The primitive data objects in DFL programs are *signals*. A signal s is an infinite stream $s = (s_1, s_2, s_3, \ldots)$ of discrete values, where each s_t denotes the value of s in sampling period t. If s is an input signal, then its values are determined by the system environment. Otherwise, they are specified by a *signal definition* of the form "s = <expression>;" where <expression> typically is an arithmetic expression on other signals. A DFL program is a collection of signal definitions, including the definition of one or more output signals. There is no notion of variables and control-flow, but DFL obeys a pure *data-flow semantics*. A signal value is defined according to its signal definition, as soon as all arguments are available. A unique definition must be present for each signal. As a consequence, the order of signal definitions in the source code is of no importance for the program semantics. DFL can be regarded as a textual format for signal flow graphs (SFGs), as introduced in section 1.4.1. An example is given in fig. 4.4. As far as data-flow descriptions are concerned, DFL is almost identical to its predecessor SILAGE. In order to make DFL applicable to a wider range of algorithms, it additionally comprises *imperative language constructs*, i.e. variables and variable assignments.

Similar to the C language, a DFL program is a set of functions, at least comprising a designated "main" function. The interface of the main function specifies input and output signals of a DSP algorithm. A DFL function consists of a set of signal declarations and signal definitions. In addition to simple signal definitions mentioned above, signals can be defined by *conditional expressions*, such as

```
s = if  a > b   -> a     /* "->" corresponds to "then" */
    || a == b  -> 0     /* "else if" part              */
    || b                /* default value               */
```

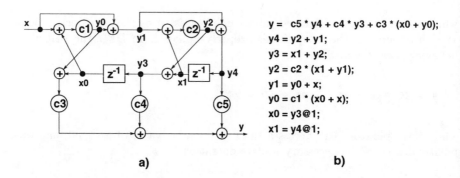

Figure 4.4 Signal flow graph (a) and DFL description (b) of 2nd order lattice filter

```
       fi;
```

Definition of signal arrays is simplified by *compound signal definitions*, most of which are *finite iterations* as in the following example:

```
signal A, B: int<10>[12]; /* two arrays of 12 10-bit integers */
(i:0..11)::              /* for all i = 0..11 */
  begin
   A[i] = x + y;
   B[i] = A[i] * 17;
  end;
```

4.6 INTERMEDIATE REPRESENTATION

The basic entities of our intermediate representation are data-flow trees. We have informally defined DFTs in section 4.2 in order to outline the difference between graph-based and tree-based code generation. In the sequel we use a more precise notation.

4.6.1 Expression tree assignments

The following definitions capture the structure of DFTs, as well as the "destinations", into which DFT are evaluated. Furthermore, different dependency relations are specified.

4.6.1.1 Definition
A **program value** V is a signal, the contents of a variable, the result of an arithmetic operation, or a constant.

A **hardware location** L is a (foreground or background) register, register file, or a primary processor port.

If L represents a register file with storage capacity greater than 1, we annotate L with a *unique symbolic address* $a \in \mathbb{N}_0$, in order to distinguish different register file elements.

A **bound program value** is a pair $B = (V, L)$, where V is a program value, and L is a hardware location.

An **expression tree node** is one of the following items:

- a **bound program value**
- a **processor input port**
- a **binary constant** $b \in \{0, 1\}^+$
- a (unary or binary) **DFL operator**
- a **cast operator** (a DFL type)
- a (ternary) **if-then-else operator**

An **expression tree** (ET) is a pair $T = (r, S)$, where the *root* r is an expression tree node, and $S = (T_1, \ldots, T_n), n \in \mathbb{N}_0$, is a (possibly empty) sequence of expression trees (*subtrees*).

An **expression tree assignment** (ETA) is a pair $A = (D, T)$, where the *destination* D is a hardware location, and T is an expression tree.

The **consumption** $C(T)$ of an expression tree T is the set of all program values, which must be defined before T can be evaluated. The **production** $P(T)$ is the program value defined by evaluation of T, i.e., by executing the computation represented by T.

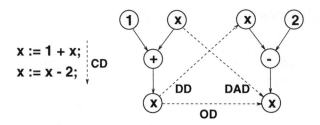

Figure 4.5 Two assignments with dependencies of types CD, DD, DAD, and OD

Let $A_1 = (D_1, T_1), A_2 = (D_2, T_2)$ be ETAs.

- A_2 is **control-dependent** on A_1 ($"A_1 \xrightarrow{CD} A_2"$), if the definition of program value $P(T_1)$ precedes the definition of $P(T_2)$ in the source code.

- A_2 is **data-dependent** on A_1 ($"A_1 \xrightarrow{DD} A_2"$), if $P(T_1) \cap C(T_2) \neq \emptyset$.

- A_2 is **data-anti-dependent** on A_1 ($"A_1 \xrightarrow{DAD} A_2"$), if $P(T_2) \cap C(T_1) = \emptyset$, and there exists a bound program value $B = (V, L)$ in T_1, such that $L = D_2$.

- A_2 is **output-dependent** on A_1 ($"A_1 \xrightarrow{OD} A_2"$), if $D_1 = D_2$ and $A_1 \xrightarrow{CD} A_2$.

The dependency relations are illustrated in fig. 4.5. In pure data-flow style DFL, only data dependencies are present. All other dependency types are meaningful only for imperative programs.

4.6.2 Background register and port binding

Before ETAs are constructed, certain program values are bound to hardware locations representing background registers. Background register binding creates a set of "anchor points" in the intermediate representation. A program value V is bound to a background register if it satisfies one of the following criteria:

1. V refers to a program variable.

Code generation

2. V is a common subexpression.
3. V refers to an array of signals or variables.
4. V is member of a delay line.
5. V is a signal with a conditional definition

If V is a primary input or output signal, it is annotated with a *port binding*, i.e. a processor port is selected, which V is assumed to be read from or written to, respectively. If a primary input signal V additionally satisfies any of the above criteria, or if multiple input signals are bound to the same processor port, then V is bound to a background register as well. We use the following strategy for background register binding[2]:

- For program values satisfying the above criteria, a *matching* background register is arbitrarily selected. A background register is matching, if it has sufficient capacity (if an array is to be bound), and its word-length is greater or equal to the word-length of the values to be bound.

- Other values (scalar signals and intermediate results) are bound to background registers only on demand, i.e. in case of resource contentions. In this case, the most appropriate background register is determined during RTP selection and RT scheduling.

4.6.3 Construction of ETAs

Basic blocks

In absence of control-flow constructs, a DFL program is translated into a basic block, consisting of ETAs. ETAs are constructed by recursive analysis of signal definitions, starting from the primary output signals. The destination for these is determined by the port binding. In the simplest case, a signal is defined by an arithmetic expression. Arithmetic expressions can be directly translated into expression trees, while inserting the necessary casts, based on the type deduction rules provided in the DFL documentation.

If the definition of a signal s_1 depends on another signal s_2, the reference to s_2 is processed as follows: If s_2 is bound to a background register R, the

---

[2] An automatic procedure for this task has recently been proposed in [SuMa95] which, however, is restricted to homogeneous memory architectures.

ETA for s_2 is constructed recursively, using register R as destination, and the reference to s_2 is replaced by the bound program value (s_2, R). The ETAs for s_1 and s_2 are marked as being data-dependent. Otherwise, the reference to s_2 is replaced by the ET constructed from the definition of s_2. In this way, ETs of maximum size are constructed within the boundaries of basic blocks and common subexpressions, independent of the structure of signal definitions.

After all ETAs have been constructed, a basic block $BB = (A_1, \ldots, A_n)$ of ETAs is computed, which obeys all inter-ETA dependencies, i.e.

$$\forall\ i, j \in \{1, \ldots, n\}:$$

$$(A_i \xrightarrow{DD} A_j \quad \vee \quad A_i \xrightarrow{DAD} A_j \quad \vee \quad A_i \xrightarrow{OD} A_j) \quad \Rightarrow \quad i < j$$

For all "meaningful" programs, the dependencies between ETAs are irreflexive and acyclic. Therefore, this type of source-level scheduling can be accomplished by computing a topological order of ETAs under the given dependencies. An example for construction of ETA basic blocks is given in fig. 4.6.

Conditionals and loops

Conditionals, such as conditional signal definitions and IF-statements, are translated into intermediate code like in classical compiler construction [ASU86]. We assume that the target machine supports an *increment operation* on the program counter PC, unconditional assignments to PC ("GOTOs"), as well as conditional assignments of the following type:

```
PC := if-then-else(<condition>, <label1>, <label2>)
```

Dependent on the value of <condition>, one of the two labels (jump addresses) is assigned to PC. We furthermore assume, that either <label1> or <label2> corresponds to an increment operation on PC, i.e. program control is passed to the "next" instruction. In the former case, an "ELSE-branch" controller is present, otherwise a "THEN-branch" controller. Which one of the two types is present in the target processor is determined by inspecting the RT patterns obtained by instruction-set extraction: Conditional assignments to the program counter correspond to guarded RTPs with destination PC and dynamic conditions.

Conditional signal definitions are translated into a sequence of control-dependent ETAs as follows: The signal is bound to a designated background register location, and separate ETAs are generated for both the THEN and ELSE part of the definition. These are augmented by ETAs to PC ensuring the correct

Code generation

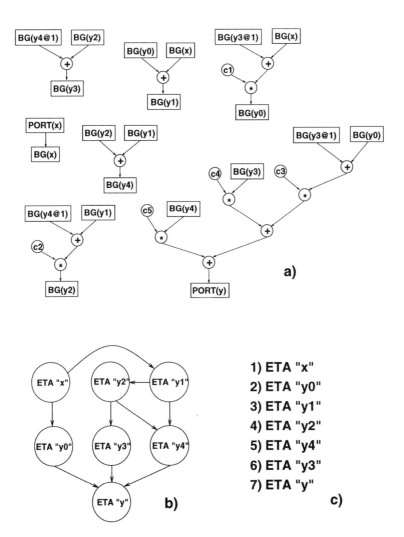

Figure 4.6 Intermediate code generation for the lattice filter example from section 4.5.3: a) Constructed ETAs, b) Data dependencies between ETAs, each denoted by its result value, c) Possible topological order. "BG(v)" denotes the background register location for program value v, "PORT(v)" denotes a port binding.

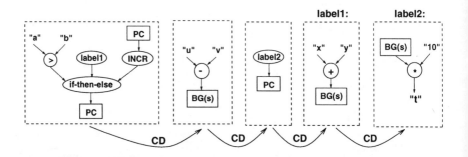

Figure 4.7 Intermediate code for conditional signal definition in presence of a "THEN-branch" controller.

control flow, while annotating ETAs with symbolic labels. References to conditionally defined signals are replaced by references to their background register locations. This is exemplified in fig. 4.7 for the following piece of DFL code:

```
s = if (a > b) -> x + y || u - v fi;
t = s * 10;
```

A similar standard mechanism is used for intermediate representation of loops. We assume, that a *reservation* (a designated memory location) exists for loop counter variables, and that the number of loop iterations is known at compile time, which is usually the case for DSP algorithms. The loop counter location is initialized with the lower loop bound. Before each execution of the loop body, the loop counter value is checked against the upper loop bound. If the upper bound is exceeded, the loop is terminated, otherwise the body is executed and the loop counter is incremented.

Many DSPs also support *hardware loops* or *zero-overhead loops* (ZOLs). In ZOLs, checking of loop boundaries is performed by special hardware, so that code overhead due to conditional branching is avoided. ZOLs are mostly realized by a dedicated *loop counter register*. The loop counter can be automatically decremented after each iteration, and the loop body is repeated until the loop counter is expired. In spite of the high importance of ZOLs for DSP code generation, here we do not focus on exploitation of ZOL hardware. This is justified by the observation that exploitation of ZOLs in *target-specific* DSP compilers is a solved problem. As indicated by results of the DSPStone project [ZVSM94], such compilers incorporate an effective macro-expansion mechanism, which recognizes FOR-loops with a fixed number of iterations and maps

Code generation   105

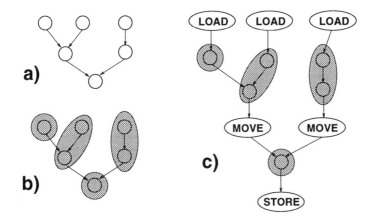

Figure 4.8 a) Sample ETA, b) result of pure covering of tree nodes, c) result of tree parsing, including load, move, and store operations

these to processor-specific ZOL hardware. Such a macro mechanism could be easily added on top of the retargetable code generation techniques presented here.

4.7 CODE SELECTION BY TREE PARSING

4.7.1 Background

Code selection for expression trees is often visualized as a process of *pattern matching* between trees and machine instructions. The goal is to compute minimum-cost *covers* for trees (fig. 4.8 a and b), i.e. the cheapest *tree implementation* by available instructions. The cost metric is given by the *number* or, more generally, by the accumulated *costs* of selected instructions. Earlier approaches solved the tree covering problem heuristically. However, particularly for inhomogeneous architectures, pure tree covering is insufficient, because covering only accounts for *operations* in ETAs, while the costs of moving values between operations are not captured. In contrast, considering the code selection problem for ETAs as a *tree parsing* problem also includes the necessary load, store, and move operations (fig. 4.8 c). The tree parsing approach makes use of the fact, that instruction sets can be written in form of *tree grammars*, a special case of context-free grammars [AhU172]. In the context of code generation, tree grammars are usually attributed with a *cost function*, which provides a quality metric for derivations. The following definitions are based on [BDB90].

4.7.1.1 Definition
A **ranked alphabet** is a finite set V together with a *ranking function* (or *arity function*) $r : V \rightarrow \mathbb{N}_0$.

Let $V_k := \{v \in V \mid r(v) = k\}$. The **tree language** $TR(V)$ over V is the smallest set, such that

- $V_0 \subseteq TR(V)$, and
- if $v \in V_k$ and $t_1, \ldots, t_k \in TR(V)$, then $v(t_1, \ldots, t_k) \in TR(V)$

A **tree grammar** is a quintuple
$$G = (\Sigma_T, \Sigma_N, S, R, c)$$
where Σ_T is a ranked alphabet of *terminals*, Σ_N is a finite set of *non-terminals* with $\Sigma_N \cap \Sigma_T = \emptyset$, $S \in \Sigma_N$ is the *start symbol*, R is a finite set of *rules*, and $c : R \rightarrow \mathbb{N}_0$ is a *cost function*. All rules $r \in R$ are of the form "$X \rightarrow t$", where $X \in \Sigma_N$, and $t \in TR(\Sigma_T \cup \Sigma_N)$ (non-terminals having rank 0).

Let $t_1, t_2 \in TR(\Sigma_T \cup \Sigma_N)$. t_1 **derives** t_2 in G, if there exists a rule $r : X \rightarrow t_3 \in R$, such that t_2 results from replacing a leaf labelled X in t_1 by t_3 (notation: "$t_1 \Rightarrow_r t_2$").

A **derivation** of a tree $t \in TR(\Sigma_T)$ in G is a sequence of rules $(r_1, \ldots, r_n)$, $n \in \mathbb{N}$, $r_i \in R$, such that there exists a sequence of trees $(t_1, \ldots, t_{n-1})$, $t_i \in TR(\Sigma_T \cup \Sigma_N)$, with
$$S \Rightarrow_{r_1} t_1 \ldots \Rightarrow_{r_{n-1}} t_{n-1} \Rightarrow_{r_n} t$$
(notation: $S \stackrel{*}{\Rightarrow} t$). The derivation has **minimum costs**, if there is no other derivation $(r'_1, \ldots, r'_{n'})$ for t in G, such that
$$\sum_{i=1}^{n} c(r_i) > \sum_{i=1}^{n'} c(r'_i)$$

The **language** of a tree grammar G is defined by
$$L(G) := \{t \mid t \in TR(\Sigma_T) \text{ and } S \stackrel{*}{\Rightarrow} t\}$$

A **tree parser** for a tree grammar G is an algorithm, which for any tree $t \in TR(V)$ for some ranked alphabet V decides whether $t \in L(G)$, and, if so, constructs a minimum cost derivation for t.

A **tree parser generator** is an algorithm, which for any tree grammar G constructs a tree parser for G.

Intuitively, terminals in a tree grammar can be associated with constants, variables, and operators, while non-terminals represent registers. Derivations in tree grammars can be graphically represented by *parse trees*. If grammar rules are identified with RT patterns, a minimum-cost parse tree reflects an optimal pattern selection for an ETA with respect to the specified cost function.

4.7.2 The iburg tree parser generator

Since the late eighties, several tree parser generators have become publicly available in the area of compiler construction [ESL89, AGT89, FHP92b]. In our approach, we use the iburg tool [FHP92a]. The iburg tree parser generator has been mainly developed at Princeton University. It accepts extended Backus-Naur form specifications of tree grammars and generates C source code for grammar-specific tree parsers. Grammar specifications for iburg are of the form shown in fig. 4.9. In the *declaration part*, the grammar start symbol and the set of terminals are defined. The declarations are followed by an arbitrary number of *rules*. Each rule is tagged with a unique rule number and an integer cost value. The trees on the right hand side of rules are either binary or unary trees, or a single terminal or non-terminal symbol.

The tree parser generated from such a specification exploits the fact, that the principle of dynamic programming can be applied to the tree parsing problem: For any input tree T, a minimum cost derivation can be obtained by computing derivations for the subtrees of T, and combining these to a derivation for T. In iburg-generated parsers, this is implemented by making two passes over the input tree T, which is illustrated in fig. 4.10.

1. A bottom-up **labelling phase**, during which each tree node is annotated with a set of matching grammar rules and the corresponding cost values. At each node n, labelling computes the minimum cost derivation for the subtree below n by dynamic programming.

```
<iburg grammar>     ::= { <declaration> } %% { <rule> }

<declaration>       ::= %start <non-terminal>
                      | %term { <terminal> = <terminal number> }

<rule>              ::= <non-terminal> : <tree> = <rule number> (<cost>);

<tree>              ::= <terminal> ( <tree> , <tree> )
                        /* binary tree rule */
                      | <terminal> ( <tree> )
                        /* unary tree rule */
                      | <terminal>
                        /* terminal rule */
                      | <non-terminal>
                        /* chain rule */

<terminal>,
<non-terminal>      ::= <character string>

<terminal number>,
<rule number>,
<cost>              ::= <integer>
```

Figure 4.9 "Meta-grammar" for specification of tree grammars in **iburg**

2. A top-down **rule emission phase**, during which the optimum parse tree is explicitly constructed based on the information of the labelling phase.

During both phases, each tree node is visited exactly once. Therefore, the runtime of tree parsing is linear in the tree size, with a factor determined by the underlying tree grammar.

Code generation

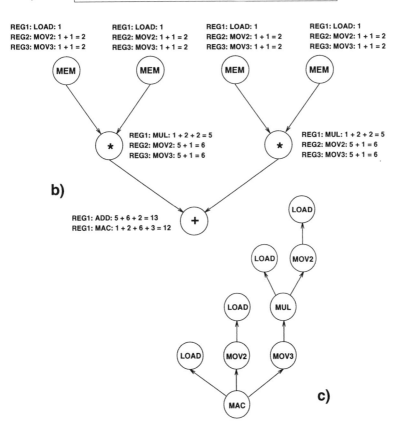

Figure 4.10 Tree parsing by dynamic programming: a) Sample tree grammar specification b) Result of labelling an expression tree: The labels "$X : Y : C$" at each tree node n denote that n can be derived from non-terminal X by rule Y at a cost of C. c) Resulting parse tree

4.7.3 Generation of tree parsers from instruction sets

This section describes, how a set $\mathcal{P}$ of RT patterns obtained by instruction-set extraction can be transformed into an iburg-compliant specification of a tree grammar $G(\mathcal{P})$. On $G(\mathcal{P})$, we can call iburg to generate a parser for ETAs with respect to the instruction set described by $\mathcal{P}$. For sake of simpler notation, we assume availability of functions $\text{TERM}(x)$ and $\text{NONTERM}(x)$, which for any "hardware entity" x (register, memory, I/O port, hardwired constant, operator) return a unique terminal resp. non-terminal symbol. Furthermore, we use the notations $REG(\mathcal{P})$, $P_{IN}(\mathcal{P})$, $P_{OUT}(\mathcal{P})$, $OP(\mathcal{P})$, and $CON(\mathcal{P})$ for the sets of all registers, processor input ports, processor output ports, operators, and (hardwired) constants that occur in RT patterns of $\mathcal{P}$.

Obviously, the terminals in $G(\mathcal{P})$ must correspond to the node types in expression trees (registers, input ports, operators, constants) and ETA destinations (output ports, registers). Therefore, the terminal set Σ_T of $G(\mathcal{P})$ is defined as

$$\begin{aligned}\Sigma_T = \;& \{\text{TERM}(x) \mid x \in REG(\mathcal{P}) \;\cup\; P_{IN}(\mathcal{P}) \;\cup\; P_{OUT}(\mathcal{P}) \\ & \cup\; OP(\mathcal{P}) \;\cup\; CON(\mathcal{P})\} \\ & \cup\; \{\text{IW}, \text{ASSIGN}, \text{SUBRANGE\_<hi>\_<lo>}, \text{BRANCH}\}\end{aligned}$$

Also the output ports $P_{OUT}(\mathcal{P})$ are represented in Σ_T, so as to match the *start rules* defined below. The designated terminal IW denotes the instruction word. Since the binary settings of instruction words are adjusted by the compiler, IW can serve to match any binary constant. The terminals ASSIGN, SUBRANGE_<hi>_<lo>, and BRANCH serve special purposes as explained later.

Non-terminals in $G(\mathcal{P})$ represent hardware objects to which values can be assigned, i.e. registers and output ports. Therefore, we define the non-terminal set Σ_N by

$$\Sigma_N = \{\text{NONTERM}(x) \mid x \in REG(\mathcal{P}) \;\cup\; P_{OUT}(\mathcal{P})\} \;\cup\; \{\text{START}\}$$

where START denotes the designated grammar start symbol.

The rule set R of $G(\mathcal{P})$ consists of four types of rules:

1. Start rules: The destination of an ETA can be any register or processor output port. Therefore, the start symbol for $G(\mathcal{P})$ must be "generic", i.e. it must permit to derive ETAs independent of their destinations. This can be achieved by introducing designated *start rules* of the form

$$\text{START} \rightarrow \text{ASSIGN}\,(\,\text{TERM}(D),\, \text{NONTERM}(D)\,)$$

Code generation

for each destination $D \in REG(\mathcal{P}) \cup P_{OUT}(\mathcal{P})$ of a GRTP contained in $\mathcal{P}$. Start rules essentially serve as "dummies", which ensure that for any ETA with destination D and having a derivation from NONTERM(D), this derivation is always found independently from D. The cost value of start rules is set to zero.

2. RT rules: RT rules immediately represent the RT patterns in $\mathcal{P}$. For any RT pattern $\mathcal{R} = (d, e)$ with destination d and RT expression e, a grammar rule "LEFT(d) $\rightarrow$ RIGHT(e)" is constructed, where

$$\text{LEFT}(d) = \text{NONTERM}(d)$$

$$\text{RIGHT}(e) = \begin{cases} \text{TERM}(e), \\ \quad \text{if } e \text{ is a binary constant} \\ \text{NONTERM}(e), \\ \quad \text{if } e \text{ is a register read access} \\ \text{TERM}(e), \\ \quad \text{if } e \text{ is a port read access} \\ \text{TERM}(op)(\text{RIGHT}(e_1)), \\ \quad \text{if } e \text{ is a unary expression } op(e_1) \\ \text{TERM}(op)(\text{RIGHT}(e_1), \text{RIGHT}(e_2)), \\ \quad \text{if } e \text{ is a binary expression } op(e_1, e_2) \\ \texttt{SUBRANGE\_<hi>\_<lo>}(\text{RIGHT}(e_1)), \\ \quad \text{if } e \text{ is a subrange expression } e_1.(< hi >:< lo >) \end{cases}$$

Since we assume single-cycle RTs, the cost value for RT rules is set to one.

3. Jump rules: A special treatment is required for RT patterns having the program counter PC as the destination. Such patterns are used to match conditional constructs in the source program. As explained in section 4.6.3, we consider conditional and unconditional jumps, as well as PC increment operations. The latter can be inserted after code compaction (cf. section 5.5.3) into all control steps that do not contain jumps. Therefore, PC increment operations do not need to be represented in the tree grammar. For RT patterns, that assign an immediate constant (an instruction word subrange) to PC, a "GOTO" rule of the form

NONTERM(PC) $\rightarrow$ `SUBRANGE_<hi>_<lo>`(IW)

for the corresponding instruction word subrange (<hi>:<lo>) is added to the grammar. Conditional assignments to PC are represented by "branch rules". Branch rules are constructed for each RT pattern $\mathcal{R} \in \mathcal{P}$ which assigns an immediate constant to PC, and which has a Boolean RT expression e as a *dynamic condition*. For these, grammar rules of the form

NONTERM(PC) → BRANCH(RIGHT(e), SUBRANGE_<hi>_<lo>(IW))

are constructed. BRANCH rules are used to match ETAs with if-then-else (ITE) tree nodes. The increment operations on PC, which, dependent on the controller type, appear in the THEN or ELSE-subtree of an ITE node, need not to be explicitly represented: Due to the assumptions made in section 4.6.3, it is guaranteed that PC is incremented if the branch is not taken. It is therefore sufficient to match only the if-then (for a "THEN-branch" controller) or the if-else part (for an "ELSE-branch" controller) of an ITE node, which avoids the necessity of ternary tree rules in the grammar. All jump rules are assigned a cost value of one.

4. Stop rules: Stop rules are additional dummy rules, which ensure context-dependent register use during tree parsing, i.e. either in terminal or non-terminal form: Consider an expression tree T with a primary input value V bound to a register file REG. A correct derivation of T is only possible, if the tree node for V is derived to terminal TERM(REG). Thus, in all RT rules comprising a read access to REG on their right hand sides, REG must appear in its terminal form. However, it might be necessary to allocate REG also for intermediate results when evaluating T. Therefore, REG *simultaneously* needs to appear in its non-terminal form NONTERM(REG), in order to permit further derivations starting from NONTERM(REG). This conflict is resolved by introducing a *stop rule* for each register of the form

NONTERM(REG) → TERM(REG)

with a cost value of zero. These stop rules permit to terminate derivations from REG, whenever primary expression tree nodes are reached during parsing.

Runtime results

Table 4.1 gives experimental results for generation of tree parsers for those target processors that have already been used for evaluation of ISE (see table 3.5 in section 3.9). Column 2 in table 4.1 shows the number of generated grammar rules for each architecture[3]. Columns 3 to 5 give the CPU seconds (measured on a SPARC-10 workstation) for the three steps of parser generation, namely grammar generation, generation of parser source code by iburg, and parser compilation, in this case by the GNU C compiler gcc. The data indicate, that

---

[3] Note, that the number of GRTPs in table 3.5 may differ from the number of grammar rules in table 4.1 because of two reasons: Firstly, the data in table 3.5 denote the number of *versions* of GRTPs, whereas a grammar rule is present only once for each pattern. Secondly, the grammars of table 4.1 contain *transformation rules*, which are explained in the following section.

Code generation

architecture	# rules	CPU grammar	CPU iburg	CPU gcc
demo	439	1.6	0.4	7.5
simplecpu	166	0.7	0.1	2.9
jmpcpu	55	0.3	0.1	1.7
ref	1703	11.7	4.1	43
manocpu	207	1.3	0.2	3.4
tanenbaum	323	1.0	0.5	8.8
simpleprocessor	32	0.2	0.1	1.1
TMS320C25	356	23	0.4	7.1
DBB	89	0.7	0.1	2.3

Table 4.1 Runtime results for tree parser generation

the goal of short turnaround times is achieved: Once a new MIMOLA processor model is developed, adaptation of the code generator to the new target (including ISE and tree parser generation) is typically possible within seconds of CPU time. For parsing expression trees of realistic size, the CPU time required by the compiled tree parsers is below the accuracy of measurement tools. Experiments have shown, that up to one thousand RT patterns are emitted per CPU second. The advantage of using such high-speed tree parsers will become obvious during discussion of transformation rules in the next section.

4.7.4 Transformation rules

Tree parsers generated as described in the previous section construct parse trees for ETAs with respect to a processor-specific tree grammar. The constructed derivations are optimal both in terms of the accumulated costs of used grammar rules and in terms of parser runtime behavior. Due to presence of register-specific rules in the tree grammar, parsing an ETA simultaneously performs RTP selection and register binding. Complex grammar rules ensure optimal exploitation of chained operations, while chain rules guarantee minimization of data moves from and to special-purpose registers. By removing zero-cost rules (start and stop rules) from a parse tree and replacing the remaining rules by their corresponding RT patterns, a **register-transfer tree** (RTT) is obtained, which represents an ETA implementation by a minimal number of available RTPs. In spite of optimality of tree parsing, the actual quality of RTTs still depends on how well ETAs are adapted to the target hardware. Two types of "distortions" between software (ETAs) and hardware (RTPs) may obstruct code selection:

Structural mismatches: The structure of arithmetic expressions in ETAs may not optimally match the hardware structure. In fig. 4.11, this is exemplified for our example target `SimpleProcessor`. Suppose, that code is to be selected for an ETA `x := a + b + c + d`, with program values `x, a, b, c, d` bound to memory `RAM.storage`. The costs of RTTs generated by parsing strongly depend on the expression tree structure. In case of `SimpleProcessor`, the left-biased structure (fig. 4.11 a) is the most appropriate. In contrast, a right-biased or a balanced structure (fig. 4.11 b and c) introduce extra stores to a temporary memory location `temp`, because of the asymmetry of RT pattern `REG.R := REG.R + RAM.storage`. On other architectures, however, the latter two ETA structures might result in cheaper RTTs.

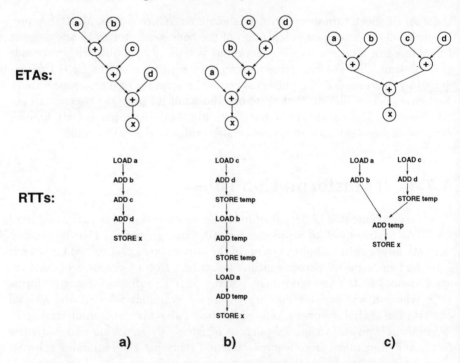

Figure 4.11 Parsing differently structured ETAs with respect to RT patterns of `SimpleProcessor`: The abbreviations LOAD, ADD, STORE denote the RTPs REG.R := RAM.storage, REG.R := REG.R + RAM.storage, RAM.storage := REG.R, respectively. a) left-biased (5 RTs), b) right-biased (9 RTs), c) balanced tree structure (7 RTs)

Code generation

Operator mismatches: Peculiarities in the target architecture may result in the fact, that there is no one-to-one mapping of ETA nodes to RTP operators. As an example, consider the TMS320C25 **SACL** instruction, which stores the lower 16 bits of the 32-bit ACCU into the 16-bit wide background register file B0. Implicitly, **SACL** involves a logical left shift (MIMOLA operator SHIFTLL) of the ACCU contents by a shift value supplied as bit index subrange (10 : 8) of the instruction word. In the TMS320C25-specific tree grammar, the rule for **SACL** therefore becomes

B0 → SHIFTLL (SUBRANGE_15_0 (ACCU), SUBRANGE_10_8 (IW))

Thus, any ETA whose derivation necessarily includes the **SACL** rule, but which does not contain an explicit shift operator node, is not accepted by the tree parser. Further operator mismatches may result from source code operators which are not present in the target hardware.

In order to circumvent such mismatches, it is necessary to apply *transformation rules* to ETAs. A transformation is a semantic-preserving mapping between ETAs. Application of transformation rules to an ETA yields a set of *alternatives*, from which the cheapest one (w.r.t. a certain cost metric) can be selected. Possible transformations comprise application of *algebraic rules* (e.g. commutativity, associativity, neutral elements of operators), while others merely serve the purpose of adaptation of ETAs in order to avoid operator mismatches. Whereas processor-specific compilers incorporate a built-in set of such transformations, based on detailed knowledge of the target architecture, a retargetable approach demands for a more flexible mechanism, which can be realized in form of a user-definable *transformation library*. Using tree parsing for code selection effectively supports the use of transformation rules because of two reasons:

1. Tree parsing for a single ETA is extremely fast, so that a comparatively large number of alternatives can be consecutively tried, from which the best one can be selected.

2. Exponential explosion of the number of alternatives due to exploitation of commutativity of operators can be avoided by inclusion of *mirror rules* in tree grammars: For each grammar rule of the form

    $X \rightarrow op(t_1, t_2)$

   with *op* being the terminal symbol of a commutative operator, the rule

    $X \rightarrow op(t_2, t_1)$

   is additionally included in the tree grammar. In this way, the tree parser is burdened with the choice from alternative ETAs arising from commutativity. Especially for DSP code generation, this feature is of outstanding importance, because the most frequent operations in DSP algorithms (addition and multiplication) are commutative. The increase in computation time for tree parsing due to mirror rules is limited by a constant factor, and is therefore negligible in the context of embedded code generation.

Our approach makes use of two types of transformation rules:

Automatic rules: During generation of tree grammars, additional rules are generated, which reflect transformations that can be selected by the tree parser. This comprises the above-mentioned mirror rules for all commutative MIMOLA operators. of *converse operators*. For instance, a "<" comparison can be replaced by a ">" comparison if the arguments are swapped. The resulting additional rules can resolve both structural and operator mismatches.

Application-specific rules: In addition to the automatically generated rules, the user may specify an external library of transformation rules specific to the intended application.

Since automatic rules do not significantly affect the compilation speed, these are included independently of the target machine. With respect to application-specific rules, we distinguish two modes: Transformation rules may express either *replacements* or *alternatives*. The former are used to *substitute* expression (sub)trees, while the latter serve the purpose of *extending the search space* of code selection. The grammar for specification of transformation libraries is shown in fig. 4.12. A transformation library consists of an arbitrary number of *rules*, each annotated with either replacement or alternative mode. A rule consists of a *typed match pattern* and a *rewrite pattern*. Since transformations may be type-specific, match patterns can be annotated with a type in order to restrict application of rules to expressions of a certain DFL data type, whereas the symbol **ANY** matches any type. A match pattern may be a formal *parameter* (line 8 in fig. 4.12), which matches any expression tree. *Typed* parameters (line 9) are used for replacement of *cast operations* in ETAs. *Binary constant* patterns (line 10) are necessary to permit matching of constants by means of instruction word index subranges instead of hardwired constants. Complex match patterns (lines 11, 12) are unary or binary tree patterns in prefix notation. These patterns contain a DFL operator and have typed match patterns as arguments in turn. Complex patterns are used for matching specific subtrees of

```
(1)   <library>              ::=   { <mode> <rule> ; }
(2)   <mode>                 ::=   REPLACEMENT
(3)                                | ALTERNATIVE
(4)   <rule>                 ::=   <typed match pattern>
                                      -> <rewrite pattern>
(5)   <typed match pattern>  ::=   ( <type> ) ( <match pattern> )
(6)   <type>                 ::=   <DFL type>
(7)                                | ANY
(8)   <match pattern>        ::=   <parameter>
(9)                                | <type> ( <parameter> )
(10)                               | <binary constant>
(11)                               | <DFL operator>
                                      ( <typed match pattern> )
(12)                               | <DFL operator>
                                      ( <typed match pattern>,
                                        <typed match pattern> )
(13)  <rewrite pattern>      ::=   <parameter>
(14)                               | IW
(15)                               | SUBRANGE ( <int>, <int>,
                                                <rewrite pattern> )
(16)                               | <binary constant>
(17)                               | <MIMOLA operator>
                                      ( <rewrite pattern> )
(18)                               | <MIMOLA operator>
                                      ( <rewrite pattern>,
                                        <rewrite pattern> )
(19)  <parameter>            ::=   <character string>
```

Figure 4.12 Specification of application-specific transformation rules

expression trees. The rewrite patterns on the right hand side of transformation rules specify those expressions, by which ETA subtrees matching the left hand side of a rule are substituted. A *parameter* rewrite pattern (line 13) generates a copy of the subtree matching the same parameter on the left hand side of the rule. The symbol **IW** denotes the instruction word, and is typically used in combination with **SUBRANGE** patterns (line 15) in order to supply arbitrary binary constants. **SUBRANGE** patterns select bit index subranges of other rewrite patterns. Further types of rewrite patterns are binary constants (line 16), or unary or binary tree patterns on MIMOLA operators.

Application of transformation rules during code selection can be implemented as follows. Given a set of application-specific transformation rules and an ETA A, a recursive tree pattern matching scheme is used to determine those rules, which match the root of the expression tree T of A, and possibly parts of its subtrees. Matching parts of T are substituted by the corresponding rewrite patterns, while retaining the original version of A in case of rules in *alternative* mode. For each generated alternative tree parsing is invoked, which includes application of automatic rules. The resulting RT tree is scheduled in order to determine the total amount of RTs including spills, as will be explained in section 4.8. For each RT tree, scheduling returns an RTL basic block. Finally, the shortest of all alternative RTL basic blocks is selected.

This transformation library concept provides a flexible mechanism for dealing with structural and operator mismatches. Typical applications of transformation rules include application of algebraic rules, exploitation of neutral elements for arithmetic operators, and implementation of data type casts in hardware. Obviously, specification of appropriate application-specific rules demands for a certain amount of "expert knowledge" to be manually encoded into the transformation library. However, since most decisions about utilization of rules are still automatically made by the code generator, the manual description effort is much lower than in the "targeted" approach discussed in section 4.3.

4.8 RT SCHEDULING

Since the tree parser has no notion of register file sizes, generated RTTs may imply *register deadlocks*, which demand for insertion of spill code. This section presents a spill mechanism suitable for resolving register deadlocks for inhomogeneous architectures.

4.8.1 Register deadlocks

RTTs are trees, in which nodes are instances of RT patterns used to implement ETAs, while edges represent data dependencies between RTs. The data dependencies in RTTs induce a *partial order*. RT scheduling is concerned with embedding such a partial order into a *total order*. The necessity of this step is a consequence of the integration of pattern selection and register binding for inhomogeneous architectures. In some cases, RTTs are degenerated (see e.g. fig. 4.11 a and b). For such linear RTTs, only data dependencies are present,

Code generation

so that a total order is already implied by code selection. In general, however, further interdependencies must be taken into account (cf. [DLSM81]):

4.8.1.1 Definition
For a register transfer x in an RTT T, the **write location** $write(x)$ denotes the destination (register or output port) of x. The **read set** $read(x)$ of x denotes the set of all registers and input ports referenced in the RT expression of x. Let x_i, x_j be different RTs of an RTT T.

- x_j is **data-dependent** on x_i ($"x_i \xrightarrow{DD} x_j"$), if $write(x_i) \in read(x_j)$, and x_i is an immediate predecessor of x_j in T. DD^* denotes the transitive closure of relation DD.

- x_j is **output-dependent** on x_i ($"x_i \xrightarrow{OD} x_j"$), if $write(x_i) = write(x_j)$ and not $x_i \xrightarrow{DD^*} x_j$.

- x_j is **data-anti-dependent** on x_i ($"x_i \xrightarrow{DAD} x_j"$), if there exists an x_k, so that $x_k \xrightarrow{DD} x_i$ and $x_k \xrightarrow{OD} x_j$.

A **(sequential) schedule** or **evaluation order** for an RTT T is a sequence (an RTL basic block) $BB = (x_1, \ldots, x_n)$ of all nodes in T.

A schedule $BB = (x_1, \ldots, x_n)$ is **valid**, if for any different RTs x_i, x_j, x_k in BB the following validity conditions hold:

$$x_i \xrightarrow{DD} x_j \Rightarrow i < j \qquad (4.1)$$

$$x_i \xrightarrow{OD} x_j \Rightarrow (i < j \ \land \ (x_i \xrightarrow{DD} x_k \Rightarrow k < j)) \ \lor$$
$$(j < i \ \land \ (x_j \xrightarrow{DD} x_k \Rightarrow k < i)) \qquad (4.2)$$

The validity conditions ensure, that values are produced, but not destroyed, before they are consumed, so that semantical correctness is preserved. Construction of valid schedules without introducing additional RTs is impossible in presence of *register deadlocks*. An RTT T has a **register deadlock**, if it contains RTs x_i, x_j, so that for any schedule one of conditions (4.1), (4.2) is violated. This situation is exposed, if tree parsing has caused a *symmetric data-anti-dependency*, which is equivalent to a "cross-wise" output dependency, as illustrated in fig. 4.13.

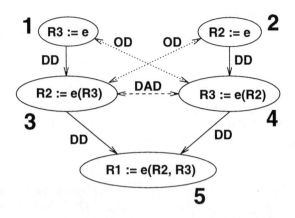

Figure 4.13 Register deadlock in an RTT T (e denotes an arbitrary RT expression): RT 3 potentially overwrites register R2 which is read by RT 4, and vice versa, so that RTs 3 and 4 are symmetrically data-anti-dependent. Thus, any evaluation order for T implies destruction of a live value.

Register deadlocks are not encountered in code generation for homogeneous architectures, because homogeneity implies that register binding can be performed *after* code selection without a loss in code quality [AhJo76, ASU86]. This is possible, because for any RTT $T = (root, (T_1, T_2))$ there is an optimal schedule with the following property: All subtrees in T, that are evaluated into memory, are scheduled first, and the rest of T is scheduled by either postorder (T_1–T_2–$root$) or anti-postorder (T_2–T_1–$root$) traversal. Such a schedule is called *contiguous*. Since evaluation of any subtree leaves at most one live foreground register, it can be ensured that this register is not used in the neighboring subtree by appropriate numbering of registers.

For inhomogeneous architectures, Araujo [ArMa95] defined the "*RTG criterion*", under which register deadlocks are excluded. The RTG criterion implies that pattern selection and register binding cannot cause a situation as shown in fig. 4.13. For architectures satisfying the RTG criterion, he proposed to perform contiguous scheduling, where the evaluation order of RTT subtrees is steered by data-anti-dependencies: If the destination register of the root of subtree T_1 also occurs as a destination in subtree T_2, then T_2 must be evaluated first, and vice versa. In this way, a valid schedule is established with the same number of RTs as in the original RTT. Architectures not satisfying the RTG criterion are, however, not considered.

If the target processor shows instruction-level parallelism, contiguous scheduling is in general not a good choice, because exploitation of parallelism may demand for an evaluation order, which "oscillates" between neighboring RTT subtrees. Non-contiguous scheduling can be performed by considering the scheduling problem as a pure code compaction problem, as in MSSQ, however at the expense of possible failure in case of register deadlocks. In order to avoid this problem, yet retaining the advantage of non-contiguous scheduling, we construct valid schedules (total RTT evaluation orders) only as an *intermediate* basic block representation, which is then passed to the compaction phase. In contrast to previous work, we do not exclude register deadlocks in advance, but resolve deadlocks by register spilling, based on heuristic spill cost estimation. The pseudo-code for RT scheduling is shown in fig. 4.14. Algorithm RTSCHEDULE takes an arbitrary RTT T and emits an RTL basic block BB while resolving possible register deadlocks. If T contains just a single RT, no scheduling is necessary (lines 7-9). Otherwise, similar to the Aho-Johnson algorithm, first all subtrees in T are determined, which are evaluated into a background register. Evaluation of these subtrees leaves no live foreground registers, so that they can be scheduled first, independently of the remaining part of T (lines 11-14). Then, the rest of T is scheduled as follows: If T is a unary tree, then its subtree T_1 is scheduled, followed by its root x (lines 16-17). If T is a binary tree with subtrees T_1, T_2, procedure SELECTSUBTREE is used to determine, whether T_1 or T_2 is to be scheduled first in order to avoid overwriting of live values (lines 18-24). If T has a register deadlock, spill code needs to be generated in order to preserve semantical correctness. Procedure INSERTSPILLS scans the basic block BB while keeping track of the number of live values in foreground registers. Whenever at a certain position in BB writing a value v into a foreground register R would exceed the capacity of R, spill code its inserted, which moves another value w currently stored in R to a background register file. Furthermore, code is generated, which reloads w into R before that position in BB, where w is read by an RT. Since in RTTs all values are consumed exactly once, the spill value w can be arbitrarily chosen.

Procedure SELECTSUBTREE (line 19) uses the following decision criteria. First, it is checked whether spilling can be avoided by applying Araujo's technique [ArMa95]. For an RTT T, we determine the set of all foreground registers, which appear as a destination in any subtree of T. This permits to identify data-anti-dependence. If for the binary input RTT $T = (x, T_1, T_2)$ the destination of the root of T_1 is not a destination in a subtree of T_2, then the live value left by evaluating T_1 cannot be destroyed by any RT in T_2, and T_1 can be scheduled first. Otherwise, the complementary test is performed between T_2 and T_1. If both tests fail, a potential register deadlock is exposed, and two heuristics are used: The first one checks, whether the destinations of the roots of T_1 and T_2

---

(1) **algorithm** RTSCHEDULE
(2) **input:** RTT T;
(3) **output:** RTL basic block BB;
(4) **var** SUB: **set of** RTTs;
(5) $\quad T'$: RTT;
(6) **begin**
(7) **if** T consists of a single RT x
(8) $\quad$ **then return** x;
(9) **end if**
(10) $BB := \emptyset$;
(11) $SUB :=$ BACKGROUNDSUBTREES(T);
(12) **for all** $T' \in SUB$ **do**
(13) $\quad BB := BB \circ$ RTSCHEDULE(T');
(14) **end for**
(15) **case** T **of**
(16) $\quad$ UNARY TREE $T = (x, T_1)$:
(17) $\quad\quad BB := BB \circ$ RTSCHEDULE(T_1) $\circ$ x;
(18) $\quad$ BINARY TREE $T = (x, T_1, T_2)$:
(19) $\quad\quad$ **if** SELECTSUBTREE(T) $= T_1$
(20) $\quad\quad\quad$ **then** $BB := BB \circ$ RTSCHEDULE(T_1) $\circ$
(21) $\quad\quad\quad\quad\quad\quad\quad$ RTSCHEDULE(T_2) $\circ$ x;
(22) $\quad\quad\quad$ **else** $BB := BB \circ$ RTSCHEDULE(T_2) $\circ$
(23) $\quad\quad\quad\quad\quad\quad\quad$ RTSCHEDULE(T_1) $\circ$ x;
(24) $\quad\quad$ **end if**
(25) **end case**
(26) $BB :=$ INSERTSPILLS(BB);
(27) **return** BB;
(28) **end algorithm**

Figure 4.14 Algorithm for RT scheduling ($\circ$ denotes concatenation)

have different capacities. In that case, the subtree with the destination having larger capacity is selected, because scheduling this subtree first probably causes less spill code.

If both capacities are equal, the spill costs for both subtrees are estimated, and the subtree with lower estimated cost is returned. Like code selection, spill cost estimation must take into account the inhomogeneity of target architectures: For special-purpose registers, different spill routes (possibly crossing other reg-

Code generation 123

isters) with different costs may exist. It is therefore reasonable to reuse the tree parser for determination of minimum-cost spill routes. This can be accomplished by constructing two new ETAs, which represent spilling and reloading the register contents to/from a background register. Code for these assignments is generated by tree parsing, which also indicates the spill costs. Note, that in spite of optimal tree parsing, the spill cost value is only an *estimation*, because if spill routes cross other registers containing live values, recursive calls of INSERTSPILLS might be necessary.

4.8.2 Example

We illustrate RT scheduling by means of an expression tree from the lattice filter example (cf. figs. 4.4 and 4.6), which is translated into TMS320C2x RTL code. Fig. 4.15 a) shows the original expression tree with primary input values c3, c4, c5 (constants) and y4, y3, y3@1, y0 (memory values). The RTT generated by tree parsing (exploiting commutativity of operators + and *) is depicted in fig. 4.15 b). Algorithm RTSCHEDULE (fig. 4.14) first schedules subtrees evaluated into background registers, in this case the subtree T^* of RTT T_{12} rooted at RT "TEMP := ACCU". This subtree leaves no live (critical) registers and cannot contribute to spill costs. For the remaining tree, it must be decided, whether T_1 or T_2 is to be scheduled first. If T_2, which leaves a live value in register PR, were scheduled first, then both T_{11} and T_{12} could not be evaluated without introducing spill code for PR. Therefore, procedure SELECTSUBTREE emits T_1, for which RT scheduling is invoked recursively. Analogously, T_{11} needs to be scheduled before T_{12}. Scheduling T_{11} leaves with a live value in ACCU, so that the remaining two RTs of T_{12} can be scheduled without conflicts (since the first three RTs of T_{12} have already been scheduled in the beginning). Finally, the root of T_1 and subtree T_2 are scheduled, followed by the root of the complete RTT. The complete sequential schedule is shown in fig. 4.15 c).

4.8.3 Mode activation

During discussion of instruction-set extraction we have seen that – besides binary encodings and dynamic conditions – execution of RTs may demand for specific setting of mode registers (cf. section 3.4.3). The register-transfer condition (RTC) extracted for each RT pattern accounts for required mode register states in form of a Boolean function. The sequential code generated by tree parsing and RT scheduling is only correct, if possibly necessary modes for a generated RT are guaranteed to be activated, before the RT is executed. The

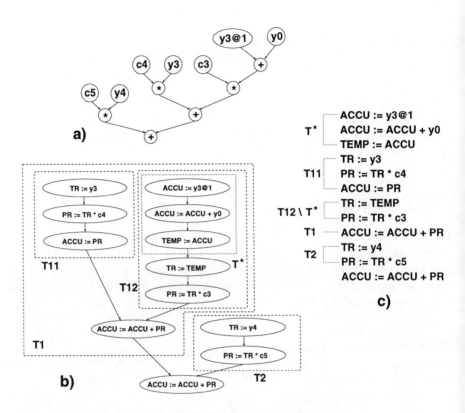

Figure 4.15 Example for RT scheduling: a) expression tree, b) register-transfer tree with subtree structure, c) spill-free sequential schedule

final step in vertical code generation therefore is insertion of additional RTs, which activate the required modes. For realistic target processors, we may make two simplifying assumptions on mode registers, which – by exploiting the generated tree parser and information obtained by ISE – reduce the task of mode activation to a relatively straightforward procedure: Firstly, mode registers are only read by module control ports, but not by any data inputs. Therefore, spilling and re-loading of mode register contents is usually neither necessary nor possible, but only "immediate load" operations on mode registers need to be considered. Secondly, we assume that the register mode requirements of a certain RT pattern do not depend on the partial instruction, which is selected during binary code generation. In order to activate the required modes, for each RTL basic block BB and for each RT x in BB, the following steps are performed:

1. Let R be the RT pattern instantiated by x, and let $F = P_1 \vee \ldots \vee P_m$ be the RT condition of R in sum-of-products form, as presented in section 3.8.2. Each product term P_j can be decomposed into $P_j = I_j \wedge M_j$, where I_j is a partial instruction, and M_j is a function on mode register bits. Since we assume mode requirements to be independent of binary encodings, $M_1 = \ldots = M_m =: M$ must hold, and F can be written as $F = (I_1 \vee \ldots \vee I_m) \wedge M$.

2. If M is a constant "true" function, then pattern R has no mode requirements. Otherwise, M is decomposed into $M = M_{r_1} \wedge \ldots \wedge M_{r_k}$ where each r_i is a mode register, and M_{r_i} is a product term on the single bits of r_i, which evaluates to "true" exactly for one binary setting b of r_i.

3. An expression tree assignment "$r_i := b$" is generated and is covered by tree parsing. Since mode registers are loaded with immediate values, the resulting register-transfer tree consists of a single RT pattern R'. An instance of R' is inserted into the vertical code immediately before RT x.

Since mode register requirements may repeat for different RTs within the same block, it is favorable to keep track of mode register states so as to avoid redundant mode activations. As an example for mode activation, consider the sequential code from the previous section in fig. 4.16. The first two RTs involve memory-to-accumulator transfers and thus an extension of the number representation of memory values y3@1 and y0. The RT conditions of the corresponding RT patterns determined during ISE therefore encode the state "1" of the TMS320C2x *sign extension mode flag* SXM. An additional RT pattern for "SXM := 1" is generated and is inserted at the beginning of the block. Furthermore, three RTs require state "0" of *product mode register* PM, which contains a shift value for PR-to-accumulator transfers. An RT pattern for "PM := 0" is generated and is inserted before the first RT requiring this mode. Further mode activations are not required or redundant, respectively.

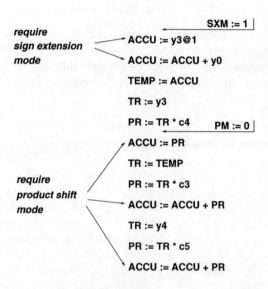

Figure 4.16 Activation of register modes in sequential schedules

5
INSTRUCTION-LEVEL PARALLELISM

Exploitation of potential parallelism is obviously a major source of code optimization. This chapter therefore focusses on DSP-specific techniques, which aim at parallelization of generated vertical machine code. In the first part, we consider the area of memory address generation. Address generation for DSPs is strongly related to instruction-level parallelism, because taking into account the DSP-specific address generation hardware permits to maximize *potential* parallelism. In the second part of this chapter, we focus on exploitation of potential parallelism by code compaction. Code compaction identifies potential parallelism, accordingly arranges RTs in time, and generates executable machine code. We analyze the special demands on compaction techniques for DSPs, and we present a novel exact solution to the problem of local code compaction.

5.1 ADDRESS GENERATION IN DSPS

The code generation techniques from the previous chapter perform binding of program values to specific foreground or background register files, but do not assign values to specific (physical) addresses within register files. In order to obtain executable machine code, the latter binding, which we call *address assignment*, has to be established. In case of foreground register files, address assignment can be regarded as non-critical, because foreground register file indices have small word-lengths, and are usually *immediate*, i.e. are encoded in the instruction word. Therefore, any address assignment for foreground registers, that does not violate the constraints imposed by the lifetimes of program values, is equivalent and has no impact on code quality. In contrast, encoding the relatively long indices for background register files (i.e. on-chip memory addresses) into the instruction word is not recommendable. Instead, DSPs provide special hardware support for address generation, which does influence code quality.

While code selection for expression trees is essentially a process of pattern matching, the task of address generation for memory accesses must rely on knowledge about operator semantics. That is, the code generator must *recognize* certain parts of a processor model as being dedicated to address generation. Doing this in a retargetable fashion is only possible, if hardware implementations of address generation units (AGUs) in DSPs do not appear in too large a variety. We therefore first analyze AGUs of several contemporary DSPs, and we identify common characteristics of these. This permits definition of a generic AGU model, which, by setting of certain parameters, matches a subset of address generation capabilities of many DSPs, and thus allows for retargetability. Based on this AGU model, several optimization goals can be defined, for which we present effective heuristic algorithms. We consider three widespread families of fixed-point DSPs, whose address generation capabilities are sketched in fig. 5.1.

TMS320C2x family

The TI TMS320C2x [TI90] supports direct and indirect memory addressing. In direct addressing mode, a 16-bit effective memory address results from concatenation of the 9-bit *data page pointer* (DP) register and the lower 7 bits of the instruction word. In indirect addressing mode, the address is supplied by an AGU containing 8 registers AR[0], ..., AR[7] of 16 bits width. The effective address is the contents of AR[ARP], i.e. the address register currently pointed to by 3-bit register ARP. Any AR can be loaded immediately, or an immediate value can be added to or subtracted from an AR register. Alternatively, address registers can be *post-modified*, i.e. can be updated at the end of a machine cycle. Post-modifications can take place in parallel to most machine instructions. Available modifications are adding or subtracting the value 1 to/from AR[ARP], or adding or subtracting the contents of AR[0] to/from AR[ARP]. Additionally, ARP can be loaded with a new immediate value.

Motorola 56xxx family

In the Motorola 56xxx [Moto92], direct addresses are supplied as immediate values in an "instruction extension" word, i.e. the word following the current instruction in program memory. The AGU for indirect addressing comprises 8 address registers R0,...,R7 and 8 *offset registers* N0,...,N7 of 16 bits width

Instruction-level parallelism

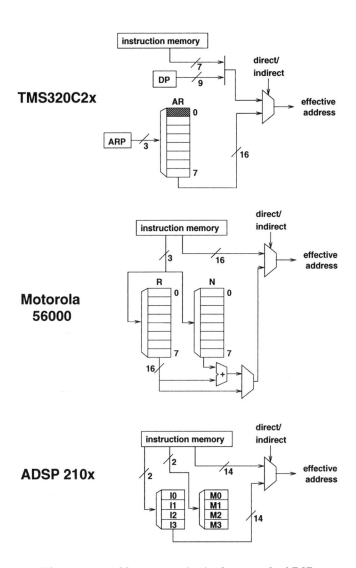

Figure 5.1 Address generation in three standard DSPs

each. The register index range 0–7 is divided into two halves, each of which is used to address one of the two memory banks "X" and "Y". An effective address is either given by the contents of register Ri, where the index i is encoded in the instruction word, or by Ri indexed by offset Ni, i.e. Ri + Ni. Both R and N registers can be loaded immediately from the instruction word. Each Ri can be post-modified either by +1/-1 or by its associated offset register Ni. Additionally, the Motorola 56xxx offers a *pre-decrement* mode, which decrements a register Ri *before* it is used for memory access. Modifications of one Ri both in lower and upper index range can take place in parallel to an arithmetic operation.

ADSP-210x family

In the Analog Devices 210x [Ana91], direct 14-bit memory addresses are supplied by the instruction word. Alternatively, indirect addressing is supported by 4 *index registers* I0,...,I3 and 4 *modify registers* M0,...,M3 of 14 bits width each, which can be loaded immediately. The effective address is given by the contents of a directly addressed index register. In parallel to each operation involving memory access by indirect addressing, one index register is post-modified by the contents of a modify register, i.e. the modify register contents are interpreted as a signed number and are added to the index register. In contrast to the Motorola, index and modify registers are not paired, but can be used orthogonally.

5.2 GENERIC AGU MODEL

We observe, that the realization of direct addressing shows significant differences, while the following features for indirect addressing are present in all of the above DSP architectures:

- A set of **address registers** (ARs), which can store effective addresses in indirect addressing mode.

- A set of **modify registers** (MRs), storing values for AR updates.

- **Auto-modify operations**, which can be executed in parallel to other machine operations, and which load an AR at the end of a machine cycle using a value supplied by the AGU itself.

Address generation by means of auto-modifys is cheaper than by immediate loads and modifys in terms of instruction count: The latter carry immediate values, which occupy a large portion of the total instruction word-length, so that these usually inhibit execution of further operations in parallel. In contrast, if next-address generation exclusively employs AGU resources, then parallelization of operations is not obstructed. Our general optimization goal

Instruction-level parallelism 131

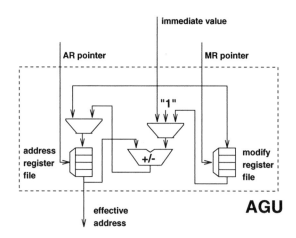

Figure 5.2 Generic address generation unit (AGU) model for DSPs

in address assignment therefore is to maximize utilization of indirect addressing and post-modify capabilities of AGUs.

In the following we consider *indirect* addressing based on the generic AGU model depicted in fig. 5.2, and we assume that such an AGU is present for each memory of the target processor. The AGU contains a file of k address registers, and a file of m modify registers. The indices for ARs and MRs are provided by two AGU inputs: the AR pointer (ARP) and the MR pointer (MRP). The third AGU input is an immediate value, originating from the instruction word, which can be used to load AR[ARP] or MR[MRP], or to immediately modify AR[ARP]. Further possible AR modifications are adding or subtracting the contents of MR[MRP] to/from AR[ARP], or adding the value +1/-1 to AR[ARP]. The detailed AGU parameters for a certain target processor can be easily extracted from the processor model and the RT pattern set.

Abstracting from the detailed AGU parameters, we consider an AGU as a black box providing a designated set of RT patterns (*"AGU operations"*) as shown in table 5.1. Their functionalities are given in C-like notation, where "imm" denotes an immediate value. Like all other RT patterns, we assume AGU operations to be executed in a single machine cycle, so that the results are valid in the following cycle.

Except for "MR load", "ARP load" and "MRP load", all AGU operations update an address register and can thus be used to generate memory addresses.

operation	functionality	cost value
"AR load"	AR[ARP] = imm	1
"MR load"	MR[MRP] = imm	1
"immediate modify"	AR[ARP] += imm	1
"auto-increment"	AR[ARP] ++	0
"auto-decrement"	AR[ARP] - -	0
"auto-modify"	AR[ARP] += MR[MRP]	0
"ARP load"	ARP = imm	0
"MRP load"	MRP = imm	0

Table 5.1   AGU operations and cost values

However, as explained above, some operations are cheaper than others, which can be expressed by introducing a relative cost metric for AGU operations. "AR load", "MR load", and "immediate modify"[1], which involve immediate values in the instruction word, cannot be performed in parallel to other operations, but introduce an extra machine instruction. Therefore, we assign the cost value 1 to these operations. On the other hand, "auto-increment", "auto-decrement", and "auto-modify" only utilize AGU resources and can be regarded as zero-cost operations. The same holds for "ARP load" and "MRP load": These require only "short" immediate values (of length 2 or 3), which are (in direct form) instruction word fields, or (in indirect form) originate from registers which can be loaded in parallel (cf. TMS320C2x). In indirect form, the required ARP contents must be prepared one machine cycle earlier than in direct form, but this has no impact on the cost metric.

5.3   ADDRESSING SCALAR VARIABLES

In order to exemplify optimization through address assignment, we consider an RTL basic block, in which a set of four symbolic memory locations ("variables")

---

[1] For sake of simplicity, we do not further consider the arithmetic interpretation of binary modify values here: If these are signed, then "immediate modify" can be used both for adding and subtracting a modify value. In case of unsigned modify values, we actually need to distinguish "immediate add" and "immediate subtract" operations. The same holds for "auto-modify" operations.

Instruction-level parallelism

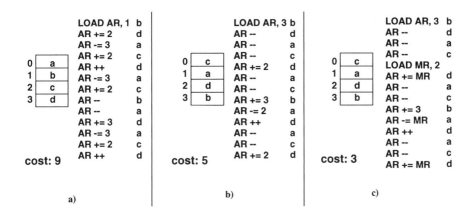

Figure 5.3 Effect of address assignment: a) "naive" assignment, b) optimized assignment, c) with utilization of modify register

$V = \{a, b, c, d\}$ are accessed in the sequence

$$S = (b, d, a, c, d, a, c, b, a, d, a, c, d)$$

Suppose, that the target DSP comprises a simple AGU with one AR, and the address space $\{0, 1, 2, 3\}$ is reserved for V. As shown in fig. 5.3, the cost of the sequence of AGU operations required to access the variables according to sequence S strongly depends on the permutation of variables in memory. Fig. 5.3 a) shows a "naive" address assignment, say

$$a \to 0, \quad b \to 1, \quad c \to 2, \quad d \to 3$$

First, AR is loaded with 1, so as to point to variable b. Then, AR is modified by +2 in order to access d, and so forth. Only 4 out of 13 AGU operations in the sequence are auto-increment/decrement operations, and the accumulated costs of the complete sequence are 9, according to the metric of table 5.1. Naive assignments are induced by an *external* ordering criterion, for instance lexicographic order of identifiers or declaration order in the source code. Address assignment exploiting access sequence information results in much lower addressing costs, as shown in fig. 5.3 b). Here, most accesses are realized via auto-increment/decrement operations, with a total cost of 5. Further reduction of addressing costs is possible, if a modify register MR is available (fig. 5.3 c): The immediate modify value "2" is needed three times in the sequence. We can load MR with 2 at a cost of 1, and reuse this value twice at zero cost. This solution implies a cost value of only 3.

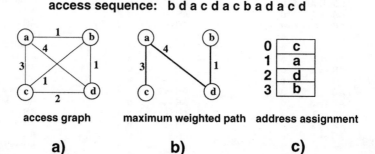

Figure 5.4 a) access graph for the example access sequence, b) maximum weighted path, c) corresponding address assignment

In the following sections we develop algorithms that construct low-cost address assignments for arbitrary variable access sequences and AGU parameters.

5.3.1 Simple offset assignment

Address assignment for DSPs has received few attention so far. Bartley [Bart92] was the first to consider the problem for the case of a single AR in detail. His algorithm is based on the observation, that two variables v_i, v_j having high *transition frequency* in the given access sequence S should be placed into consecutive memory cells, in order to enable zero-cost auto-increment/decrement addressing. The transition frequency is the number of (v_i, v_j) and (v_j, v_i) subsequences in S. Bartley proposed to represent the assignment problem by means of an *access graph*.

5.3.1.1 Definition
Given a set V of variables and a sequence S of accesses to elements of V, the **access graph** is a weighted undirected graph $G = (V, E, w)$, where E consists of all edges $\{v_i, v_j\}$, $v_i, v_j \in V$, and $w : E \rightarrow \mathbb{N}_0$ specifies the transition frequency for each $\{v_i, v_j\}$.

Fig. 5.4 a) shows the access graph for our above example. By means of access graphs, one can define the *Simple Offset Assignment* (SOA) problem as follows[2]:

---
[2] The name SOA has been introduced by Liao [LDK+95a], and is used here as well for sake of consistency.

5.3.1.2 Definition

For an access sequence S on variable set V, an **address assignment** is a mapping
$$\pi : V \to \{0, \ldots, |V| - 1\}.$$
which assigns all variables in V to a unique location within a contiguous address space of size $|V|$. The **distance** $\delta_\pi(v_i, v_j)$ of two variables $v_i, v_j \in V$ with respect to π is $|\pi(v_i) - \pi(v_j)|$.

Let $G = (V, E, w)$ be the access graph for S. The **cost** of an address assignment π is defined as
$$cost(\pi) = 1 + \sum_{e=\{v_i,v_j\}\in \hat{E}} w(e)$$
with
$$\hat{E} = \{\{v_i, v_j\} \in E \ | \ \delta_\pi(v_i, v_j) > 1\}$$

Simple Offset Assignment is the problem of computing a minimum cost address assignment for an access graph G in presence of a single address register.

There is a one-to-one mapping between variable permutations and *Hamiltonian paths* (touching each graph node exactly once) in an access graph, which is induced by defining variables with distance 1 as being adjacent nodes in the path. The cost of an address assignment is the number of access transitions between variables that are *not* assigned to consecutive addresses (plus "1" for the necessary AR initialization). Therefore, a minimum cost address assignment corresponds to a maximum weighted Hamiltonian path in the access graph (fig. 5.4 b). An optimum address assignment is given by traversing the path (fig. 5.4 c), where the direction is of no importance. Computing a maximum weighted Hamiltonian path for an access graph is NP-hard, even though access graphs are complete graphs[3]. Therefore, heuristics should be used for non-trivial SOA problems.

Liao's SOA algorithm [LDK+95a] is a derivative of Kruskal's spanning tree algorithm for graphs [HoSa87]. It first sorts all non-zero edges in G in descending order of weight and starts with an empty path P. In each step, the next *valid* edge is taken from the sorted list. An edge is valid, if (as in Kruskal's algorithm) it does not cause a cycle in P, and additionally does not cause any node in P to have a fanout larger than 2. Valid edges are added to P, until P contains $|V| - 1$ edges. If less than $|V| - 1$ non-zero edges are present, then

---
[3] This follows from a simple polynomial-time reduction from the "standard" Hamiltonian path decision problem [GaJo79].

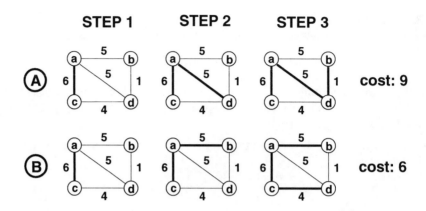

Figure 5.5 Impact of equally weighted access graph edges on SOA costs: In step 1, edge $\{a,c\}$ must be selected in any case. In step 2, there is a choice between $\{a,d\}$ and $\{a,b\}$ Choosing $\{a,d\}$ (A) enforces selection of 1-weighted edge $\{b,d\}$ in step 3, because $\{c,d\}$ and $\{a,b\}$ are invalid. If $\{a,b\}$ were chosen instead (B), then edge $\{c,d\}$ is still valid, resulting in lower SOA cost.

P is padded with zero-edges. The validity conditions for edge selection ensure, that finally P is Hamiltonian.

In the following, we present a heuristic improvement for this procedure and empirically prove its efficacy. In general, access graphs may contain many edges of identical weights. Whenever two equally weighted edges e_1, e_2 are valid at a certain point of time during SOA, arbitrary selection of either e_1 or e_2 is not a good choice, as demonstrated in fig. 5.5. Instead, one can heuristically exploit access graph information to break ties, so as to reduce SOA costs. Generalizing the example from fig. 5.5, we observe that for equally weighted edges e_1, e_2 priority should be given to that edge, which leaves the highest amount of freedom in terms of remaining valid edges to select from. However, not only the number, but also the weight of remaining valid edges should be taken into account. Since either e_1 or e_2 is selected before further edges can be classified as valid or invalid, a probability metric is required for tie-breaking. Such a metric is induced by the following "tie-break" function:

5.3.1.3 Definition
For an access graph $G = (V, E, w)$, the **tie-break function** $T_G : E \to \mathbb{N}_0$ is defined by
$$T_G(e) = \sum_{e' \in E} w(e'), \quad \text{with} \quad e \cap e' \neq \emptyset$$

Instruction-level parallelism

---

(1) **algorithm** SOLVESOA
(2) **input:** access sequence S on variable set V;
(3) **output:** address assignment π;
(4) **var** P: path in G;
(5) $\quad$ G: access graph;
(6) $\quad$ L: list of edges;
(7) **begin**
(8) $\quad$ $G = (V, E, w) :=$ access graph for S;
(9) $\quad$ $L :=$ sorted list of non-zero edges in G,
(10) $\quad\quad$ using order $e_1 < e_2$, if $w(e_1) = w(e_2)$ and $T_G(e_1) < T_G(e_2)$;
(11) $\quad$ $P := \emptyset$;
(12) $\quad$ **while** $|P| < |V| - 1$ **do**
(13) $\quad\quad$ $e :=$ next edge in L;
(14) $\quad\quad$ **if** VALID(e) /* e causes neither cycles nor trees in P */
(15) $\quad\quad\quad$ **then** INSERT(P, e);
(16) $\quad\quad$ **end if**
(17) $\quad$ **end while**
(18) $\quad$ **return** TRAVERSE(P);
(19) **end algorithm**

Figure 5.6 Improved Simple Offset Assignment algorithm

Function T_G accumulates the edge weights in the vicinity of e. For an edge e, a high T_G value indicates high relative probability, that including neighboring edges of e in the Hamiltonian path would be favorable. Selection of e restricts the number of valid edges in its neighborhood. For equally weighted edges e_1, e_2, we thus give priority to e_1, exactly if $T_G(e_1) < T_G(e_2)$. Recalling the example from fig. 5.5, we see that this heuristic actually detects the better solution, since

$$T_G(\{a,b\}) = 5 + 6 + 5 + 1 = 17 \quad < \quad T_G(\{a,d\}) = 5 + 5 + 6 + 4 + 1 = 21$$

The complete SOA algorithm, which combines Liao's technique and the tie-break function is shown in fig. 5.6. Starting with an empty path P, valid edges are added to P, until $|P| = |V| - 1$. The validity test can be efficiently implemented by a "union-find" data structure [HoSa87]. If necessary, P is completed by zero-edges. Finally, the address assignment is constructed by traversing P.

$|V|$	$|S|$	naive (100 %)	Liao	%	with tie-break	%
5	10	4.85	2.22	46	2.18	45
5	20	8.65	5.30	61	5.26	61
15	20	13.38	6.59	49	6.09	46
10	50	30.86	21.80	71	21.30	69
20	50	35.83	22.79	64	21.58	60
40	50	37.82	19.56	52	17.87	47
10	100	60.86	48.04	79	47.73	78
50	100	76.41	46.91	61	43.78	57
80	100	77.67	40.72	52	36.86	47
50	200	152.32	113.92	75	109.89	72
100	200	156.61	98.74	63	91.04	58
			average	61		58

Table 5.2 Experimental results for Simple Offset Assignment

Table 5.2 provides empirical data, which demonstrate the efficacy of tie-breaking. In order to obtain unbiased and processor-independent results, the experiments have been performed on sets of pseudo-random variable access sequences. Problem parameters are the number $|V|$ of different variables and the access sequence length $|S|$. For each pair of parameters, the *average address assignment costs* over 100 random sequences have been measured. Column 3 shows the cost values for a naive assignment, induced by lexicographic order of variables. The cost values for Liao's heuristic with arbitrary tie-breaking are given in column 4, while column 5 lists the relative costs compared to naive assignment. Columns 6 and 7 show the corresponding data when using function T_G for tie-breaking. The CPU times for both algorithms are almost identical and (on a SPARC-10 workstation) do not exceed a few milliseconds. Our SOA algorithm on average yields additional 3 % cost savings compared to naive assignment, and there is no set of parameters, for which tie-breaking leads to higher average costs than the original procedure.

5.3.2 General offset assignment

SOA only deals with address assignment in presence of a single address register, and is therefore over-simplified. DSPs typically have 4 to 8 ARs, and exploiting all of these can further reduce address assignment costs. Nevertheless, efficient SOA algorithms may serve as subroutines for solving realistic address assignment problems.

By partitioning the variable set V into disjoint subsets and assigning each subset to one of the k ARs, the above SOA procedure can be immediately used also in case of $k > 1$ ARs. However, this partitioning must be performed carefully. This leads to the definition of the *General Offset Assignment* (GOA) problem.

5.3.2.1 Definition
Given an access graph $G = (V, E, w)$ for an access sequence S, and a variable subset $V' \subseteq V$, the **subsequence** $S(V')$ induced by V' is the sequence of all accesses in S exclusively referring to elements of V'.

For an AGU with $k \in \mathbb{N}$ address registers and an access sequence S on variable set V, **General Offset Assignment** is the problem of computing a partitioning

$$P : V \to \{V_1, \ldots, V_k\}$$

so that

$$\sum_{i=1}^{k} cost(\pi_i) \to \min$$

where π_i denotes an optimal address assignment for subsequence $S(V_i)$.

As a generalization of SOA, GOA is obviously NP-hard. Liao [LDK+95a] proposed the following recursive partitioning algorithm: Starting with an access sequence S on variable set V and k ARs, a variable subset $V' \subset V$ is determined. Then, the subsequences $S(V')$ and $S(V \setminus V')$ and the corresponding access graphs are constructed. If the sum of SOA costs for $S(V')$ and $S(V \setminus V')$ are lower than the SOA costs for the original sequence S, then subset V' is "accepted", i.e. it is assigned to the k-th AR, and the algorithm is called recursively for $S(V \setminus V')$, V', and $k - 1$. Otherwise the algorithm terminates. If $k = 1$, then all remaining variables are assigned to one AR, and recursion stops. The actual address assignment is then constructed by solving SOA for the subsequence of each accepted variable subset and concatenating the resulting address assignments. Since SOA only produces memory addresses starting from 0, a specific "offset" of $|V_1|+\ldots+|V_{i-1}|$ has to be added to all addresses that refer to a certain subset V_i.

Unfortunately, Liao did not provide a general heuristic for determining the subsets V', which is a crucial step of the procedure, but he suggested to select subsets of fixed size s (typically 2 to 6), based on a certain edge-weight criterion. However, this may lead to badly balanced partitionings with $k-1$ subsets of size s, and one large subset of size $|V|-s\cdot(k-1)$. Furthermore, Liao observed that the most appropriate value for s strongly depends on the given access sequence, so that manual parameterization for each problem instance is required.

These problems can be avoided by permitting flexible subset sizes. Our GOA algorithm determines an individual variable subset size for each AR, using SOLVESOA as a subroutine for cost estimations. We start with l, $l \leq k$, subsets $V_1, \ldots, V_l$ of size two, by selecting the l disjoint edges of highest weight in the access graph and assigning the corresponding node pairs to one subset each. This ensures, that these edges do not contribute to the total GOA cost. If the access graph contains k or more disjoint non-zero edges, then $l = k$. Otherwise, all of these edges are taken, and $l < k$. The initial SOA cost for each subset is one. Then, all remaining variables v are step-by-step assigned to that subset V^*, for which the increase in SOA cost caused by adding v to V^* is minimal. After all variables have been assigned, the actual address assignment is obtained by solving l SOA problems. Its cost value is the sum of the l SOA cost values. The pseudo-code for the GOA procedure is shown in fig. 5.7.

Empirical performance results are listed in table 5.3. Columns 1 to 3 show the problem parameters. Experiments have been performed for the same $|V|/|S|$ combinations as in table 5.2 with different numbers of address registers. Column 4 shows the average (over 100 random access sequences) GOA cost values obtained with Liao's algorithm, where variable subset size s was set to 3. The GOA costs achieved by algorithm SOLVEGOA are listed in column 6, while column 8 shows the cost reduction relative to Liao's algorithm. The required computation times (CPU seconds on a SPARC-20) are given in columns 5 and 7. On the average, our algorithm yields address assignments of 22 % lower cost, and − except for parameter set (40,50,8) − does not perform worse than Liao's. Furthermore, SOLVEGOA is faster in most cases, and it does not rely on user-specified parameters, since both the number of used ARs and the subset of variables assigned to each AR are determined within the algorithm.

Instruction-level parallelism 141

```
(1)  algorithm SOLVEGOA
(2)  input: access sequence S on variable set V;
(3)     number k of ARs;
(4)  output: address assignment π;
(5)  var V₁,...,Vₖ: set of variables;
(6)     L: list of edges;
(7)     G: access graph;
(8)  begin
(9)     G = (V, E, w) := access graph for S;
(10)    L := sorted list of non-zero edges in E;
(11)    V₁,...,Vₖ := ∅;
(12)    i := 0;
(13)    repeat
(14)       i := i + 1;
(15)       {u, v} := next edge in L with u, v ∉ V₁ ∪ ... ∪ Vₖ;
(16)       Vᵢ := {u, v};
(17)    until (i = k) or ({u, v} = ∅);
(18)    l := i;
(19)    for all v ∈ V, v ∉ V₁ ∪ ... ∪ Vₗ do
(20)       V* := the element of {V₁,...,Vₗ}, for which
(21)          cost(SOLVESOA(S(V* ∪ {v}))) −
(22)          cost(SOLVESOA(S(V*))) → min;
(23)       V* := V* ∪ {v};
(24)    end for
(25)    return SOLVESOA(S(V₁)) ∘ ... ∘ SOLVESOA(S(Vₗ));
(26) end algorithm
```

Figure 5.7 Algorithm for General Offset Assignment

5.3.3 Modify registers

Accessing a sequence of variables according to the address assignment produced by SOLVEGOA can be implemented by exclusively using AGU operations on address registers. Construction of the corresponding sequence AS of AGU operations is straightforward, once the address assignment is known. In section 5.3 (fig. 5.3), we have already demonstrated, how exploitation of available modify registers can lead to further reduction of addressing costs. The general idea is to keep multiply required immediate modify values for ARs in MRs, from which modify values can be retrieved at zero cost by "auto-modify" operations.

$|V|$	$|S|$	k	Liao	CPU sec	SolveGOA	CPU sec	gain (%)
5	10	2	2.10	0.2	2.02	0.01	4
5	10	4	2.17	0.3	2.06	0.01	5
5	20	2	3.56	0.3	2.54	0.01	29
5	20	4	3.56	0.4	2.57	0.01	28
15	20	2	4.47	0.2	3.81	0.05	15
15	20	4	4.14	0.5	4.13	0.06	0
10	50	2	13.14	0.2	11.93	0.03	9
10	50	4	9.30	0.6	5.02	0.03	46
20	50	2	15.82	0.2	14.44	0.08	9
20	50	4	10.96	0.8	7.76	0.1	29
20	50	8	9.83	1.3	8.05	0.08	18
40	50	2	14.55	0.3	12.64	0.5	13
40	50	4	10.40	0.8	7.65	0.4	26
40	50	8	8.14	1.4	8.22	0.5	-1
10	100	2	30.51	0.2	27.50	0.03	10
10	100	4	17.16	0.7	8.45	0.02	51
10	100	8	17.29	0.8	5.00	0.01	71
50	100	2	39.80	0.5	34.71	0.8	13
50	100	4	30.79	1.2	23.44	0.6	24
50	100	8	21.34	2.5	12.69	0.6	41
80	100	2	34.25	0.7	28.75	6.3	16
80	100	4	27.05	1.6	19.17	4.2	29
80	100	8	17.80	3.1	11.30	3.7	37
50	200	2	100.52	0.7	92.85	0.6	8
50	200	4	81.34	1.6	67.23	0.5	17
50	200	8	56.86	3.1	38.15	0.6	33
100	200	2	89.49	1.5	78.97	8.9	12
100	200	4	77.20	3.1	60.95	5.0	21
100	200	8	59.61	5.8	37.84	3.2	37
						average	22

Table 5.3 Experimental results for General Offset Assignment

Whenever the number of modify values with multiple occurrence in AS exceeds the number m of available MRs, the decision which of these values should be assigned to MRs is non-trivial. In this section we present an efficient algorithm

Instruction-level parallelism

for optimal utilization of m MRs for a (previously generated) sequence AS of AGU operations.

For this purpose, it is sufficient to consider the subsequence AS' of AS, which consists of all "immediate modify" operations in AS with a total cost of $|AS'|$. Each of these is annotated with a modify value u, so that AS' is characterized by a sequence $U = (u_1, \ldots, u_n)$ of integer values. For instance, the AGU operation sequences from figs. 5.3 a) and b) are represented by

$$U = (2, -3, 2, -3, 2, 3, -3, 2) \quad \text{and} \quad U = (2, 3, -2, 2)$$

respectively. Similar to address assignment, we first consider the case, that only a single modify register MR is available. The *cost* $c(u_i)$ of an integer value u_i in U is the cost of the AGU operation used to generate u_i. Without usage of MR, $c(u_i) = 1$ for each u_i. When using MR, the following alternatives are present:

A1: u_i is retained as an immediate value, so that $c(u_i) = 1$.

A2: u_i is retrieved from MR by "auto-modify", so that $c(u_i) = 0$, if MR already contains u_i.

A3: u_i is first loaded into MR and is then retrieved from MR, so that $c(u_i) = 1$.

Optimal MR utilization is achieved, if the alternatives are chosen in such a way, that the sum over all $c(u_i)$ is minimized. We use the notation $next(u_i)$ to denote the index of the next occurrence of value u_i in the subsequence $(u_{i+1}, \ldots, u_n)$ of U ($next(u_i) = \infty$ denotes no further occurrence). One can easily show, that MR utilization is optimal, if for each u_i in U the alternative is chosen as follows:

1. A1, if $MR = x$, x is undefined or such that $next(x) = \infty$, and $next(u_i) = \infty$

2. A1, if $MR = x \neq u_i$, $next(x) \neq \infty$, $next(u_i) \neq \infty$, and $next(x) < next(u_i)$

3. A2, if $MR = u_i$

4. A3, if $MR = x$, x is undefined or such that $next(x) = \infty$, and $next(u_i) \neq \infty$

5. A3, if $MR = x \neq u_i$, $next(x) \neq \infty$, $next(u_i) \neq \infty$, and $next(x) > next(u_i)$

This provides criteria, for which u_i values MR usage is favorable. In case that $m > 1$ MRs are available on the target machine, it must be additionally decided, *which* of these should be used for a certain modify value u_i. If some $MR_j, j \in \{1, \ldots, m\}$ carries an undefined value, or a value x with $next(x) = \infty$, then overwriting MR_j by u_i cannot cause any disadvantage. In contrast, if all MRs carry modify values with further occurrences, then unfavorable selection of a certain MR_j may induce a number of superfluous "MR load" operations.

The problem of MR selection is equivalent to the problem of *page replacement in operating systems*: By identifying MRs with page frames, and modify values with pages, we can apply algorithms from operating system theory. As a special constraint, the complete sequence U of modify values is known in advance, which permits utilization of Belady's algorithm [Bela66]. Belady showed, that if the complete page access sequence is known, overwriting the frame containing the page with the *largest forward access distance* is optimal, i.e. leads to the smallest total number of page replacements. Analogously, in case a new modify value has to be loaded into the MR file, that MR_j must be selected, which carries the value with the maximum next-occurrence index. Whenever assigning a modify value to an MR is favorable, MR_j is guaranteed to be the optimal one. This allows to combine the above observations and Belady's algorithm to the procedure shown in fig. 5.8, which achieves optimal utilization of m MRs for a sequence AS of AGU operations. In the presented form, algorithm MR-OPTIMIZE works for AGUs with *signed* modify arithmetic and *orthogonal* AR/MR files. In case of unsigned arithmetic (as in the TMS320C2x), only the absolute values of modify values have to be considered, which leads to a higher potential of reusing MR contents. For non-orthogonal AR/MR files (as in the M56000), MR optimization must be restricted to available AR/MR index pairs. The corresponding adaptations of MR-OPTIMIZE are straightforward.

The worst-case runtime behavior of MR-OPTIMIZE is $\mathcal{O}(m \cdot n^2)$, because for each op_i all MRs have to be tested, and the rest sequence $(op_{i+1}, \ldots, op_n)$ has to be traversed in order to determine the $next()$ values. Typically, the number of multiple occurrences of u_i can be regarded as being bounded by some small constant $c \ll |AS|$, so that by storing all next-occurrence indices in advance the runtime can be reduced to $\mathcal{O}(m \cdot n)$. In practice this means, that even larger problems ($m = 8, n = 200$) are solved within 50 milliseconds on a SPARC-10. Extensive experimentation showed, that performing MR optimization on the AGU operation sequences obtained by GOA on the average reduces the addressing costs by 20 %.

The algorithms for optimized address assignment, as presented in the previous sections, work for single basic blocks. In presence of multiple basic blocks,

```
(1)  algorithm MR-OPTIMIZE
(2)  input: sequence AS = (op_1,...,op_n) of AGU operations
(3)     without MR usage; number m of MRs;
(5)  output: sequence MS of AGU operations with MR usage;
(5)  var mr: array [1..m] of integer; /* MR states */
(6)  begin
(7)      MS := ∅;
(8)      mr[1],...,mr[m] := "undefined";
(9)      for i = 1 to n do
(10)         case op_i of
(11)             IMMEDIATE MODIFY OPERATION "AR_l = u_i":
(12)                 if  ∃mr[j], j ∈ {1,...,m} : mr[j] = u_i
(13)                 then /* alternative A2 */
(14)                     MS := MS o "AR_l += MR_j";
(15)                 elseif  ∃mr[j], j ∈ {1,...,m}:
(16)                         (mr[j] = "undefined") ∨ (next(mr[j]) = ∞)
(17)                 then begin /* alternative A3 */
(18)                     MS := MS o "MR_j = u_i" o "AR_l += MR_j";
(19)                     mr[j] := u_i;
(20)                 end
(21)                 else begin /* select A1 or A3 */
(22)                     j_max := max{next(mr[1]),...,next(mr[m])};
(23)                     if next(u_i) < next(mr[j_max])
(24)                     then begin /* alternative A3 */
(25)                         MS := MS o "MR_{j_max} = u_i" o "AR_l += MR_{j_max}";
(26)                         mr[j_max] := u_i;
(27)                     end
(28)                     else MS := MS o op_i; /* alternative A1 */
(29)                 end if
(30)                 end
(31)             OTHER:
(32)                 MS := MS o op_i;
(33)         end case
(34)     end for
(35)     return MS;
(36) end algorithm
```

Figure 5.8 Algorithm for optimal MR utilization

performing address assignment separately for each block obviously may result in contradicting address assignments, because the access transitions between variables may be different for each block. We solve this problem by constructing a *global* access graph, which results from merging all local access graphs for basic blocks. Performing GOA on the global access graph thus takes address assignment requirements for different blocks simultaneously into account. If a basic block is part of a loop body, we multiply the edge weights in its access graph by the number of loop iterations, which is known at compile time. In this way, higher priority is given to "critical" pieces of code.

5.4 ARRAYS AND DELAY LINES

5.4.1 Address generation for array references

Since most of the execution time of programs is typically spent in loops, optimization of machine code generated for loops has been subject to intensive research in area of compiler construction. Different standard loop optimization techniques are described in [ASU86]. Presumably the most obvious optimization is *loop-invariant code motion*. A computation is loop-invariant, if its result is independent of values generated within a loop, so that it can be moved "outside" of the loop. Another standard technique is *induction variable elimination*. Induction variables are those variables, which are increased or decreased by the same constant within each loop iteration. Elimination of such variables may lead to *strength reduction*, that is, a – potentially costly – multiplication is replaced by an addition. Further loop optimizations include loop folding and unrolling, which were already mentioned in section 1.6.3.

The above-mentioned techniques operate at the source level and thus do hardly take into account the underlying hardware for address generation. Therefore, they mainly are useful for optimizations prior to code generation. DSP-specific techniques can make use of the fact, that array traversal in DSP programs tends to follow a very regular, mostly linear scheme. Therefore, it is justified to consider a restricted version of the problem of address generation for array accesses. Here, we make the following simplifying assumptions.

- We consider FOR-loops with a *fixed number of iterations*. The value range of loop variables must be known at compile time.

- For each loop nesting level, there is a *unique loop variable*, which is initialized before the first iteration starts, and which is incremented at the end of each iteration. Array index expressions must only refer to such loop variables.

- We consider array references with simple array index expressions. If
$$A[f_1(i_1)][f_2(i_2)]\ldots[f_n(i_n)]$$
denotes a reference to an n-dimensional array A, then all indices $f_j(i_j)$ must be an expression of the form $"c"$, $"i_j+c"$, $"i_j-c"$, or $"c-i_j"$, where c is a constant in $\mathbb{N}_0$, and i_j is a loop variable.

In the following we consider, how array address computations satisfying the above assumptions can be efficiently mapped to DSP AGUs. Like for the offset assignment problems SOA and GOA, our goal is to maximize utilization of auto-increment/decrement capabilities. In Araujo's approach [ASM96], a similar graph-based problem formulation as the one presented here is defined. Optimization of address generation, however, is restricted to single loop iterations. Liem [LPJ96] has presented a tool that transforms C code, so as to maximize increment/decrement operations on pointers. Its effectiveness has been demonstrated for SGS Thomson DSP cores. However, no concrete algorithms have been published.

We first examine simple (non-nested) loops. A simple loop is a finite iteration over a basic block of the form

```
FOR i = lo TO hi DO
   array_ref_1
   array_ref_2
   ...
   array_ref_k
END
```

The basic block, which forms the loop body, is given as an RTL basic block, i.e. vertical code generated by RT scheduling (section 4.8). Here, we only consider the sequence of array references within the loop body. The loop variable, say i, as well as its lower and upper bound are specified in the DFL source code. After each iteration, variable i is implicitly incremented by 1. Each array reference array_ref_i is of the form

$$A[f_1(i)]\ldots[f_n(i)]$$

where A denotes the array identifier, n is the array dimension, and $f_1,\ldots,f_n$ denote array index computations. Since we are dealing with non-nested loops with a single loop variable i, exactly one of the $f_j(i)$ expressions is assumed to be non-constant. Array references with purely constant indices can be regarded as a special case of accesses to scalar values.

As in standard compiler construction [ASU86], we use a line-oriented assignment of array elements to memory cells. The general address mapping for an element $A[m_1]\ldots[m_n]$, $m_i \in [0, N_i - 1]$, of an $N_1 \times \ldots \times N_n$ array A is given by (cf. [HoSa87]):

$$adr(A[m_1]\ldots[m_n]) = base(A) + \sum_{j=1}^{n} m_j \cdot a_j$$

with

$$a_j = \begin{cases} \prod_{k=j+1}^{n} N_k, & 1 \leq j < n \\ 1, & j = n \end{cases}$$

Each array reference can be associated with an integer *update value*. Let i_0 and i_1 denote the values of loop variable i in two consecutive loop iterations. The update value UV is defined as the difference between the referenced memory addresses of two consecutive loop iterations:

$$UV(A[f_1(i)]\ldots[f_n(i)]) =$$
$$adr(A[f_1(i_1)]\ldots[f_n(i_1)]) - adr(A[f_1(i_0)]\ldots[f_n(i_0)])$$

If the address register used for a certain array reference is properly initialized, then UV denotes the value, which has to be added to that address register in order to prepare the same array reference for the next iteration. The above assumptions ensure, that all update values are constants and can be computed at compile time. Our goal is to enable efficient generation of required update values by AGU resources. As we will see in the following example, however, manual identification of such good address generation schemes can be a time-consuming task already for small loop bodies. Instead, a proper assignment of array references to address registers needs to be systematically computed, based on the update values and the number of available address registers.

It is easily observed, that two array references can only share an address register, if both have the same update value. Otherwise, no consistent machine code could be generated[4]. Sharing an address register between two array references is useful, if the difference of the memory addresses referenced within a single loop iteration has an absolute value less or equal to 1, because in this case our

---

[4]Since the binary encoding of the address register pointer value is part of a machine instruction, switching the address register between loop iterations would imply to modify the encoding of this instruction at runtime.

AGU model permits to generate one address from another at zero cost. We illustrate this for the following loop:

```
FOR i = 2 TO N DO
   ref A[i+1]    (1)
   ref A[i]      (2)
   ref A[i+2]    (3)
   ref A[i-1]    (4)
   ref A[i+1]    (5)
   ref A[i]      (6)
   ref A[i-2]    (7)
END
```

In this example, all references are of the form "$A[i+c]$", thus having an update value of 1. Therefore, an address generation scheme with a single address register, say AR1, could be used. This would induce the following AGU operation sequence for the loop:

```
AR1 = adr(A[3])      // point to initial address of A[i+1]
FOR i = 2 TO N DO
   (1) AR1 --        // point to A[i+1], prepare (2)
   (2) AR1 += 2      // point to A[i], prepare (3)
   (3) AR1 -= 3      // point to A[i+2], prepare (4)
   (4) AR1 += 2      // point to A[i-1], prepare (5)
   (5) AR1 --        // point to A[i+1], prepare (6)
   (6) AR1 -= 2      // point to A[i], prepare (7)
   (7) AR1 += 4      // point to A[i-2],
                     //    prepare (1) for next iteration
END
```

Only two out of seven AGU operations in the loop body are (zero cost) autodecrements. The alternative "extreme" solution would be to reserve a separate address register for each reference, which would avoid any addressing overhead, except for more initializations. However, this would restrict the optimization potential of address generation for further data objects. Therefore, we must assign array references to a number of address registers, which in general is lower than the number of references. This can be achieved based on the following graph model.

5.4.1.1 Definition
For a simple loop with sequential array references $(r_1, \ldots, r_k)$, the **intra-iteration distance graph** G_{intra} is a directed acyclic graph (V, E), with

$V = \{r_1, \ldots, r_k\}$. There is an edge $e = (r_i, r_j) \in E$, exactly if $i < j$, $UV(r_i) = UV(r_j)$, and $|adr(r_i) - adr(r_j)| \leq 1$.

Presence of an edge $e = (r_i, r_j)$ in G_{intra} reflects the fact, that the address for r_j can be computed from $adr(r_i)$ at zero cost. Thus, r_i, r_j might share an address register without inducing overhead. Fig. 5.9 shows the intra-iteration distance graph for the above example. However, G_{intra} does not capture address relations between array references across the boundaries of a single loop iteration. Relationships between different iterations are considered in the following definition.

5.4.1.2 Definition

For a simple loop with sequential array references $(r_1, \ldots, r_k)$, the **inter-iteration distance graph** G_{inter} is a bipartite directed acyclic graph (V, E), with $V = \{r_1, \ldots, r_k\} \dot\cup \{r'_1, \ldots, r'_k\}$ and $E \subseteq \{r_1, \ldots, r_k\} \times \{r'_1, \ldots, r'_k\}$. The nodes $\{r'_1, \ldots, r'_k\}$ represent the references $\{r_1, \ldots, r_k\}$ in the next loop iteration. There is an edge $e = (r_i, r'_j) \in E$, exactly if $i \geq j$, $UV(r_i) = UV(r_j)$, and $|adr(r_i) + UV(r_i) - adr(r_j)| \leq 1$.

Presence of an edge $e = (r_i, r'_j)$ in G_{inter} reflects the fact, that the address for reference r_i in the *following* loop iteration, which is equal to $adr(r_i) + UV(r_i)$, can be computed from $adr(r_j)$ at zero cost. The inter-iteration distance graph for the example loop is shown in fig. 5.10. In order to simultaneously represent intra-iteration and inter-iteration dependencies, we merge both graph models.

5.4.1.3 Definition

For a simple loop with an intra-iteration distance graph $G_{intra} = (V_1, E_1)$ and an inter-iteration distance graph $G_{inter} = (V_2, E_2)$, the **overall distance graph** (ODG) is the directed acyclic graph $G_{dist} = (V_1 \cup V_2, E_1 \cup E_2)$.

Given a loop with array references $(r_1, \ldots, r_k)$, assigning the references to address registers means to partition the set $\{r_1, \ldots, r_k\}$ into a number of disjoint groups, which is less or equal to the number of available address registers. According to the definition of the ODG, the addresses for a group of references can be generated at zero cost (neglecting the address register initialization), whenever the ODG comprises a path, that touches all nodes in that group. For a certain array reference r_i, the address register (the register number, not

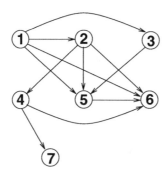

Figure 5.9 Intra-iteration distance graph for the example loop: Graph edges represent opportunities for address register sharing. For instance, starting from the address of reference (1), one can use auto-decrement to obtain the address of reference (2). Alternatively, the address for reference (3) could be generated by auto-increment. Another choice would be to keep the address of (1), which can be reused for reference (5).

its contents) used to provide the address for r_i must be the same for all loop iterations. Otherwise, self-modifying code would be necessary. Thus, if r_i, r_j, $i < j$, are selected to share an address register, and no reference r_l, $l < i$ is also assigned to that register, then there must exist a path $(r_j, \ldots, r'_i)$ in the ODG.

In total, optimally assigning array references to address registers results in a *path covering problem* of the following form:

5.4.1.4 Definition
For a simple loop with an ODG $G_{dist} = (V, E)$, $V = \{r_1, \ldots, r_k, r'_1, \ldots, r'_k\}$, the problem of **address register assignment** is to compute a minimal number of node-disjoint paths $(P_1, \ldots, P_m)$ in G_{dist} (with $P_i = (p_{i1}, \ldots, p_{in_i})$, $p_{ij} \in V$), such that each path starts at a node $r_i \in \{r_1, \ldots, r_k\}$, each path ends in the corresponding node $r'_i \in \{r'_1, \ldots, r'_k\}$, and all nodes in $\{r_1, \ldots, r_k\}$ are touched by exactly one path.

Since this is an intractable problem[5], exact solutions cannot be computed for loop bodies with a larger amount of array references. Araujo [ASM96] proposed to compute an exact solution for a restricted problem definition, which neglects inter-iteration distances. If – in our formulation – only the intra-iteration dependence graph G_{intra} is considered, then a linear-time procedure can be used

---
[5] A special case is to decide existence of two node-disjoint paths with given start and end nodes in a DAG, which is shown to be NP-complete in [RoSe90].

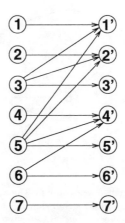

Figure 5.10 Inter-iteration distance graph for the example loop: Since all update values are equal to 1, all edges $e = (r_i, r'_i)$ are trivially part of the edge set E. Reference (1) (A[i+1]) becomes "A[i+2]" in the next iteration. Therefore, its address can generated by keeping the address of reference (3), or by auto-increment of the address for reference (5). Similarly, the address for reference (3) could be auto-decremented, so as to provide the address for reference (2) in the next iteration.

to compute an optimal path cover [BoGi77]. However, a potential overhead is caused by the necessity to update address registers for the "next" loop iteration. In order to avoid this potential overhead, we use a heuristic algorithm, which does take inter-iteration distances into account. The idea is to iteratively compute *longest paths* in the ODG, which obey the restrictions from definition 5.4.1.4. Since the ODG is acyclic, this can be accomplished by breadth-first traversal. First, the longest path P in the complete ODG is determined, and the elements of P are assigned to the first address register. Then, all nodes in P are deleted from the ODG, as well as their incident edges. On the remaining graph, the longest path is computed in turn, and its nodes are assigned to the next address register. This is iterated, until all ODG nodes are covered by a path.

The pseudo-code for heuristic address register assignment is given in fig. 5.11. While the presented algorithm aims at minimizing the number of address registers, resource limitations might impose an upper bound M of address registers that are available for array addressing. Algorithm ASSIGN ADDRESS REGISTERS comprises a simple method to ensure, that this restriction is not violated. If only a single address register is left (lines 16–17), then all remaining array ref-

Instruction-level parallelism 153

```
(1)  algorithm ASSIGNADDRESSREGISTERS
(2)  input: simple loop L with array references (r₁,...,rₖ)
(3)     M: integer /* upper bound of number of address registers */
(4)  output: partition of {r₁,...,rₖ} into m ≤ M groups
(5)  var G_intra, G_inter, ODG: DAG;
(6)     regnum: integer;
(7)     g₁,...,g_M: set of array references;
(8)     P: set of array references;
(9)  begin
(10)    regnum := 1;
(11)    g₁,...,g_M := ∅;
(12)    G_intra := BUILDINTRAITERATIONDISTANCEGRAPH(r₁,...,rₖ);
(13)    G_inter := BUILDINTERITERATIONDISTANCEGRAPH(r₁,...,rₖ);
(14)    ODG = (V,E) := MERGE(G_intra, G_inter);
(15)    while V ≠ ∅ do
(16)       if regnum = M
(17)          then P := V; /* register limit reached */
(18)          else P := NODES(LONGESTPATH(ODG));
(19)       end if
(20)       V := V \ P;
(21)       assign node set P to group g_regnum;
(22)       regnum := regnum + 1;
(23)    end while
(24)    return {g₁,...,g_M};
(25) end algorithm
```

Figure 5.11   Address register assignment for simple loops

erences are assigned to this register, regardless of address distance information. Alternatively, techniques for address register spilling could be included.

We exemplify the use of algorithm ASSIGNADDRESSREGISTERS for our example loop. Merging the graphs from figs. 5.9 and 5.10 yields an ODG with three valid longest paths, each of length four:

$$P_1 = (1, 2, 5, 1'), \quad P_2 = (1, 3, 5, 1'), \quad P_3 = (2, 5, 6, 2')$$

If we select P_1, then the remaining ODG contains the nodes $\{3, 4, 6, 7\}$. For this rest graph, there is a unique longest path (4,6,4'), and nodes 3 and 7 remain to be covered. Since neither 3' is reachable from 7, nor 7' from 3, the two trivial paths (3, 3') and (7, 7') must be selected in order to obtain a complete cover. The partitioning is thus given by

$$g_1 = \{1, 2, 5\}, \quad g_2 = \{4, 6\}, \quad g_3 = \{3\}, \quad g_4 = \{7\}$$

This requires four address registers. The AGU operation sequence for this partitioning is

```
       AR1 = adr(A[3])      // point to initial address of (1)
       AR2 = adr(A[1])      // point to initial address of (4)
       AR3 = adr(A[4])      // point to initial address of (3)
       AR4 = adr(A[0])      // point to initial address of (7)
       FOR i = 2 TO N DO
   (1)   AR1 --             // point to A[i+1], prepare (2)
   (2)   AR1 ++             // point to A[i], prepare (5)
   (3)   AR3 ++             // point to A[i+2], prepare (3)
                            //    for next iteration
   (4)   AR2 ++             // point to A[i-1], prepare (6)
   (5)   AR1                // point to A[i+1], keep address for (1)
                            //    for next iteration
   (6)   AR2                // point to A[i], keep address for (4)
                            //    for next iteration
   (7)   AR4 ++             // point to A[i-2], prepare (7)
                            //    for next iteration
       END
```

Only zero-cost AGU operations are necessary in the loop body. Note, that also selection of different longest paths would not have resulted in a solution with fewer address registers, so that the heuristic computes an optimal solution for this example. For less than four address registers there exists no address register assignment without overhead due to costly AGU operations, i.e. involving modify values different from +1/-1. For this example, one can show[6] that neglecting the inter-iteration dependencies, as discussed above, would yield a solution with two address registers, however at the expense of two costly AGU operations to be executed at the end of each loop body.

The above address register assignment technique for simple loops can be generalized for nested loops. The basic difference is, that more restrictions need to be obeyed concerning sharing of address registers. In order to exclude invalid address register sharing, we construct a separate ODG for each loop level, and

---

[6] An optimal cover for the graph from fig. 5.9 would consist of the paths $P_1 = (1, 2, 4, 7)$ and $P_2 = (3, 5, 6)$ in this case.

we demand that an ODG edge between two nodes r_i, r_j must only be present, if r_i, r_j have pairwise identical update values for all loop levels. Then, for each level, we proceed as in fig. 5.11. However, in case of deeply nested loops with a large number of array references, it might be necessary to include mechanisms for address register spilling.

5.4.2 Address generation for delay lines

As already mentioned in section 1.4.1, DSP algorithms often contain *delay lines* as a special data structure. For a given period t, a delay line is characterized by a sequence

$$D_x = (x(t), x(t-1), x(t-2), \ldots, x(t-N))$$

of signal values, where $x(t)$ denotes the value of a signal x during the t-th execution (*period*) of an infinite-loop DSP program, and $x(t-i)$, $1 \leq i \leq N$ represents the value of x from the i-th previous period relative to t.

From a code generation viewpoint, delay lines pose two problems. Firstly, delay lines need to be identified in the source code of a DSP program. For C program sources, this is a difficult task, because delay lines may be syntactically specified in numerous different ways. For the DFL language, the problem of recognizing delay lines in the source code is not present, because these are explicitly labelled by means of the DFL operator "@" (cf. section 4.5.2). For each period t, the expression "$x@i$" denotes a reference to the physical location (memory address) $adr(x(t-i))$ of $x(t-i)$. The "@" operator permits to easily identify accesses to elements of a delay line D_x as well as the length $N+1$ of D_x.

The second problem is to generate an appropriate memory organization for delay lines. At the end of each period, is has to be ensured that each value $x(t-i)$, $0 \leq i \leq N-1$, is accessible as $x(t-(i+1))$ for the next period, and the oldest element $x(t-N)$ is dropped from the delay line. This is visualized in fig. 5.12, presuming delay lines are assigned to contiguous address spaces in memory.

A naive way of updating delay lines is to use a fixed mapping of delay line elements to memory cells, and, for $i = N-1\ldots 0$, to execute data moves $adr(x(t-i)) \rightarrow adr(x(t-(i+1)))$. This overhead can be avoided if *circular buffers* or *ring buffers* are used. The basic idea in ring buffers is a *dynamic* mapping of delay line elements to memory cells. Instead of moving values between memory cells, only the address mapping adr is changed, which is

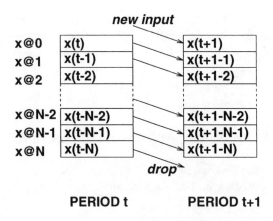

Figure 5.12   Update of delay lines in memory

accomplished by updating a single address pointer after each period. Unfortunately, an "ideal" ring buffer is usually not present in processor hardware, but realistic memories show a linear rather than a circular address space. As a compromise, many DSPs therefore provide special hardware support for *simulated* ring buffers. In the following we examine the two most widespread hardware solutions, and we describe how our generic AGU model presented in section 5.2 can be extended to provide support for delay lines.

Implementation by modulo addressing

Ring buffer simulation by *modulo addressing* is found in Motorola and Analog Devices DSPs [Moto92, Ana91]. The basic idea is to perform next-address computation using a dedicated *modulus logic*, which ensures that address registers are modified only in such a way, that the address range allocated to a ring buffer cannot be left. Fig. 5.13 shows the AGU model from section 5.2 extended by modulo addressing hardware. Addresses computed by the adder/subtracter of the AGU are passed to the modulus logic, as well as a buffer length value retrieved from a file of *length registers*. Length registers are addressed by a *length register pointer*. The modulus logic checks, whether the computed address is within the address range of the ring buffer, whose length is specified by the length register contents. If the next address exceeds the buffer range, it is "wrapped" around the buffer boundaries.

In the original AGU model from section 5.2, any next address A' was computed from a "current" address A (stored in an address register) and a modify value

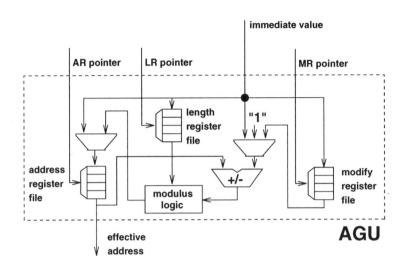

Figure 5.13 AGU with buffer length registers and modulus logic

M (representing an immediate modify value, modify register contents, or +1/-1 for auto-increment/decrement) by the simple equation

$$A' = A + M$$

For AGUs with modulo addressing capabilities, this next-address equation is replaced[7] by

$$A' = ((A + M - B) \text{ MOD } L) + B$$

where B denotes the ring buffer base address, and L is a buffer length stored in a length register. If the current address A is within the buffer range ($A \in [B, B + L - 1]$) and the modify value M satisfies $-L < M < L$ (otherwise the next address would wrap around the buffer boundaries more than once), then all addresses "$X := A + M - B$" must satisfy $X \in [-(L-1), 2(L-1)]$. For a given L, operator "MOD" implements wrapping of addresses around buffer boundaries, and is defined by:

$$X \text{ MOD } L = \begin{cases} X, & \text{if } 0 \leq X < L \\ |L - X|, & \text{if } L \leq X \leq 2(L-1) \\ L - |X|, & \text{if } -(L-1) \leq X < 0 \end{cases}$$

---

[7] For data structures other than delay lines, modulo addressing needs to be disabled by loading a designated, processor-dependent "neutral element" $\mathcal{N}$ into a length register. For instance, $\mathcal{N} = 0$ for Analog Devices and $\mathcal{N} = -1$ for Motorola DSPs. Any expression "$((A + M - B) \text{ MOD } \mathcal{N}) + B$" evaluates to $A + M$ in the modulus logic, so that modulo addressing is turned into regular linear addressing.

Since X MOD $L \in [0, L-1]$ for any X, all next addresses A' are guaranteed to be within the buffer range as well.

In the form of the above next-address equation, however, modulo addressing would imply too much overhead, because two additions and one subtraction need to be computed. Furthermore, a separate base register would be necessary. Therefore, realistic DSPs impose a restriction on ring buffer base addresses: For a ring buffer of length L, the base address B must be of the form $k \cdot 2^n$, $n, k \in \mathbb{N}$, so that $2^{n-1} < L \leq 2^n$. That is, the $\lceil \log L \rceil$ least significant bits of the base address need to be zero. All values in $[0, L-1]$ fit into these $\lceil \log L \rceil$ bits. Therefore, any address A with a binary representation $A = (a_m, \ldots, a_0)$ can be uniquely decomposed into a base

$$B(A) = (a_m, \ldots, a_{\lceil \log L \rceil})$$

and an offset

$$O(A) = (a_{\lceil \log L \rceil - 1}, \ldots, a_0)$$

with $A = B(A) \circ O(A)$ ($\circ$ denotes concatenation, which does not require extra hardware). Then, the next-address computation is reduced to

$$A' = B(A) \circ ((O(A) + M) \text{ MOD } L)$$

which avoids costly subtraction and addition of the buffer base B. A separate base register is not required, because – for given L – the base is implicit in each ring buffer address.

In our approach, we use the following procedure to implement delays lines by modulo addressing ring buffers:

1. For each delay line $D_x = (x(t), x(t-1), x(t-2), \ldots, x(t-N))$ of length $L = N+1$, an address register/length register pair AR/LR is reserved.

2. An address range $R = [B, B+L-1]$ in the memory assigned to D_x is allocated. The lowest possible range (i.e. a range not already allocated for other data) is chosen, which satisfies $B = k \cdot 2^n$ with $2^{n-1} < L \leq 2^n$.

3. The length register LR is initialized with L, and address register AR is initialized with an arbitrary address within R, for instance B.

4. Before each access to an element of D_x, an AGU operation is inserted, which executes the necessary AR modification. In presence of modulo addressing it is not required that D_x is traversed in strict order of descending or ascending delays, but an arbitrary sequence $(x@i_1, \ldots, x@i_s)$,

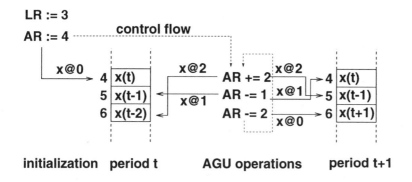

Figure 5.14 Ring buffer implementation by modulo addressing

$j \in \mathbb{N}, i_j \in [0, N]$ of accesses to elements of D_x can be implemented. The first access $x@i_1$ is prepared by AR initialization. For any next access $x@i_j$, the difference $d = i_j - i_{j-1}$ is added to AR, where modulo addressing (enabled by LR initialization) ensures that the range R is not left.

5. After the last access $x@i_s$, the value $d = i_1 - i_s - 1$ is added to AR, whereby AR is prepared to point to $x(t - i_1 - 1)$ for the next period.

Example

As an example for modulo addressing, consider a delay line $D_x = (x(t), x(t-1), x(t-2))$ of length $L = 3$. Since $2^1 < L \le 2^2$, the ring buffer base address in memory must have two zero bits at the least significant positions. Suppose, the address range $R = [4, 6]$ is chosen for D_x, and the sequence of accesses is $(x@0, x@2, x@1)$. Fig. 5.14 shows the organization of address generation for D_x. A length register LR is initialized with 3, and an address register AR is loaded with the buffer base $B = 4$. In the first period, the address mapping is

$$x@0 \leftrightarrow 4, \quad x@1 \leftrightarrow 5, \quad x@2 \leftrightarrow 6$$

After the last access $(x@1)$ AR is updated by -2, whereby the mapping for the next period becomes

$$x@1 \leftrightarrow 4, \quad x@2 \leftrightarrow 5, \quad x@0 \leftrightarrow 6$$

and AR points to the new location of $x@0$ for the next period. The modulus logic ensures, that all further AR updates correctly wrap around the buffer range $R = [4, 6]$.

Implementation by parallel data moves

Due to the additional silicon area requirements for modulo addressing hardware, the TMS320C2x DSPs show a more restricted hardware support for delay lines in the form of *parallel data moves*. Such a data move copies the contents of a memory cell $M[a]$ into the next higher cell $M[a+1]$. Data moves can be executed in parallel to certain other operations. In contrast to modulo addressing, base addresses are independent of buffer lengths when using parallel data moves. On the other hand, parallel data moves are only effective, if a delay line element $x(t-i)$ is moved exactly at that point of time when it is read, because only in this case, the address generated for reading $x(t-i)$ can be simultaneously used to move $x(t-i)$ one memory cell higher. This results in the restriction, that no random access to delay line elements is permitted, because the value of $x(t-(i+1))$ must not be overwritten by a data move, before the last access (within one period) to $x(t-(i+1))$ has taken place. We therefore demand, that the sequence of accesses to delay line elements satisfies the following precondition: For each $i \in [0, N-1]$, there must exist an access to $x(t-i)$, which follows the last access to $x(t-(i+1))$ in the sequence. In this way, a data move $adr(x(t-i)) \rightarrow adr(x(t-(i+1)))$ can be scheduled "safely" at that point of time, when the last access to $x(t-i)$ takes place. Taking these observations into account, we use the following procedure for delay line implementation by parallel data moves:

1. For each delay line $D_x = (x(t), x(t-1), x(t-2), \ldots, x(t-N))$ of length $L = N+1$, one address register AR is reserved.

2. An address range $R = [B, B+L-1]$ in the memory assigned to D_x is allocated, which is not occupied by other data. The mapping of delay line elements to memory cells is statically fixed, namely
$$\forall i \in [0, N]: \quad x@i \quad \leftrightarrow \quad B+i$$

3. Let $S = (x@i_1, \ldots, x@i_s)$ be the sequence of accesses to elements of D_x. Address register AR is initialized with $B + i_1$, so as to point to $x@i_1$ in memory.

4. For any next access $x@i_j$, the difference $d = i_j - i_{j-1}$ is added to AR, before the access to $x(t-i_j)$ takes place. If $x@i_j$ denotes the last access to $x(t-i_j)$, then a data move $adr(x(t-i_j)) \rightarrow adr(x(t-(i_j+1)))$ is additionally inserted, so that the delay line is properly updated for the next period.

5. After the last access $x@i_s$, AR is adjusted to point to $x@i_1$ for the next period by adding $d = i_1 - i_s$ to AR.

Example

We illustrate delay line implementation by parallel data moves for the previous example. Suppose, that the address range $R = [4,6]$ is reserved for $D_x = (x(t), x(t-1), x(t-2))$, and the access sequence is $S = (x@2, x@1, x@0)$. Fig. 5.15 shows the organization of address generation for D_x. Address register

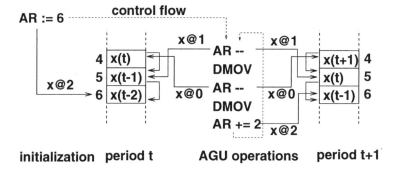

Figure 5.15 Ring buffer implementation by parallel data moves

AR is initialized with address 6 of $x@2$. The access to $x@1$ is prepared by decrementing AR followed by a data move (DMOV). The data move copies the memory contents pointed to by AR to the next higher cell number (AR)+1, i.e. $x(t-2)$ is overwritten by $x(t-1)$. The access to $x@0$ is prepared analogously. After the access to $x@0$, AR points to the base address $B = 4$, and adding 2 to AR ensures that AR points to the former $x@1$ value, which is accessed as $x@2$ in the next period. Obviously, the buffer range $R = [4,6]$ is never left.

5.5 CODE COMPACTION

The above address generation techniques only operate on the *vertical* code, while actual exploitation of *parallelism* has to be ensured by code compaction, which is the subject of this section. The code compaction procedure no longer makes a distinction between RTs generated for expression trees and AGU operations, but operates exclusively on the basis of inter-RT dependencies and instruction encodings for RT patterns.

5.5.1 Background

Code compaction is the task of parallelizing a set of RTs under a set of given constraints by assigning the RTs to *control steps*. All RTs assigned to the same control step form a microinstruction. The optimization goal in general is to achieve the minimum number of control steps. *Local* code compaction starts from an RTL basic block $BB = (x_1, \ldots, x_n)$ generated by previous compilation phases. In order to preserve semantical correctness for the compacted machine code, BB must satisfy the additional assumption that control flow "enters" BB always at RT x_1, i.e. x_1 is the only jump target in BB.

The RTs $x_1, \ldots, x_n$ show data-dependencies, data-anti-dependencies, and output dependencies, as defined in section 4.8 in the context of RT scheduling. The interpretation of relations DAD and OD for RTL basic blocks is, however, slightly different for compaction: Since RT scheduling possibly inserts spill code in order to remove symmetric data-anti-dependencies, the evaluation order of output-dependent RTs imposed by BB must not be changed, i.e. $x_i \xrightarrow{OD} x_j$, if x_i, x_j have the same destination and $i < j$, which makes OD an asymmetric relation. In turn, this implies $x_i \xrightarrow{DAD} x_j$ only for $i < j$. Therefore, $DD \cup OD \cup DAD$ is a partial ordering on $\{x_1, \ldots, x_n\}$, which can be represented by a DAG:

5.5.1.1 Definition
For an RTL basic block $BB = (x_1, \ldots, x_n)$, the **RT dependency graph** (RDG) is an edge-labelled directed acyclic graph $G = (V, E, w)$, with $V = \{x_1, \ldots, x_n\}$, $E \subseteq V \times V$, and $w : E \to \{DD, DAD, OD\}$.

In our machine model, all RTs are single-cycle operations, all registers permit at most one write access per cycle, and all registers can be written and read within the same cycle. This leads to the following "basic" definition of the code compaction problem:

5.5.1.2 Definition
A (**parallel**) **schedule** for an RDG $G = (V, E, w)$ is a mapping

$$CS : \{x_1, \ldots, x_n\} \to \mathbb{N}$$

from RTs to control steps so that for all $x_i, x_j \in V$:

$$x_i \xrightarrow{DD} x_j \quad \Rightarrow \quad CS(x_i) < CS(x_j)$$

Instruction-level parallelism

$$x_i \xrightarrow{OD} x_j \Rightarrow CS(x_i) < CS(x_j)$$

$$x_i \xrightarrow{DAD} x_j \Rightarrow CS(x_i) \leq CS(x_j)$$

Code compaction is the problem of constructing a schedule CS such that

$$\max\{CS(x_1), \ldots, CS(x_n)\} \rightarrow \min$$

The following notions are important in the context of code compaction:

The **as-soon-as-possible** time $ASAP(x_i)$ of an RT x_i is defined as (with $\max(\emptyset) := 1$):

$$ASAP(x_i) = \max(\quad \max\{ASAP(x_j) + 1 \quad | \quad x_j \xrightarrow{DD} x_i \quad \vee \quad x_j \xrightarrow{OD} x_i\},$$
$$\max\{ASAP(x_j) \quad | \quad x_j \xrightarrow{DAD} x_i\})$$

The **critical path length** L_c of an RDG is $\max_{i=1..n}(ASAP(x_i))$, which provides a lower bound on the minimum schedule length.

The **as-late-as-possible** time $ALAP(x_i)$ of an RT x_i is defined as (with $\min(\emptyset) := L_c$):

$$ALAP(x_i) = \min(\quad \min\{ALAP(x_j) - 1 \quad | \quad x_i \xrightarrow{DD} x_j \quad \vee \quad x_i \xrightarrow{OD} x_j\},$$
$$\min\{ALAP(x_j) \quad | \quad x_i \xrightarrow{DAD} x_j\})$$

An RT x_i lies on a **critical path** P_c in an RDG, if $ASAP(x_i) = ALAP(x_i)$.

In case of unlimited hardware resources, code compaction can be efficiently solved by topological sorting. Real target architectures, however, impose resource limitations, which may inhibit parallel execution of pairwise independent RTs. These limitations can be captured by an *incompatibility relation*

$$\not\parallel \quad = \quad \{(x_i, x_j) \in V \times V \quad | \quad x_i, x_j \text{ cannot be executed in parallel}\}$$

Incompatibilities impose the additional constraint

$$\forall x_i, x_j \in V : \quad x_i \not\parallel x_j \quad \Rightarrow \quad CS(x_i) \neq CS(x_j)$$

on code compaction, in which case compaction becomes a resource-constrained scheduling problem, known to be NP-hard [GaJo79].

Heuristic code compaction techniques became subject of research with appearance of VLIW machines in the early eighties. An important contribution to this area is the survey and empirical comparison by Davidson et al. [DLSM81], in which three $\mathcal{O}(n^2)$ compaction algorithms are evaluated, namely *first-come-first-served, critical path,* and *list scheduling*. Davidson et al. concluded, that each of these heuristics produces close-to-optimum results in most cases, while differing in speed and simplicity. Nevertheless, the above techniques were essentially developed for horizontal machines with few restrictions imposed by the instruction format, i.e. resource conflicts are mainly caused by restricted *data-path* resources.

5.5.2 DSP-specific requirements

Frequently, DSPs do not show horizontal, but *strongly encoded* instruction formats, in order to limit the silicon area requirements of on-chip program memories. An instruction format is strongly encoded, if the instruction word-length is small compared to the total number of control lines for RTL processor components. For such formats, most resource conflicts actually arise from *encoding conflicts*, and instruction-level parallelism in DSPs is restricted to a set of "critical cases", which are regarded to be important for DSP applications. Thus, compaction algorithms for DSPs have to scan a relatively large search space in order to detect sets of RTs qualified for parallelization.

In contrast to horizontal machines, code compaction for DSPs must take into account the following special requirements:

Conflict representation: As pointed out by Marwedel and Nowak [NoMa89], an elegant method of checking both for data-path resource conflicts and encoding conflicts in a uniform way is to use the partial instructions of RTs for conflict detection. Two partial instructions

$$I_1 = (a_1, \ldots, a_m), \quad I_2 = (b_1, \ldots, b_m), \quad a_i, b_i \in \{0, 1, x\}$$

are conflicting if

$$\exists i \in \{1, \ldots, m\}: \quad (a_i = 1 \ \land \ b_i = 0) \ \lor \ (a_i = 0 \ \land \ b_i = 1)$$

In the BDD-based notation introduced in chapter 3, this is equivalent to $F_1 \land F_2 \equiv 0$, if F_1, F_2 denote the Boolean functions for I_1, I_2. Encoding conflicts are obviously represented in the partial instructions. The same holds for data-path resource conflicts, if control codes for data-path resources are assumed to be adjusted by the instruction word[8].

Alternative versions: As pointed out in chapter 3, each RT x_i may have a set of alternative versions (partial instructions) $B_i = \{b_{i1}, \ldots, b_{im_i}\}$. Obviously, compatibility of RTs depends on the selected versions, that is, parallelizing a set of RTs may be excluded, even if all RTs are pairwise compatible. The number of alternative versions tends to grow with the degree of instruction encoding, because in encoded formats the number of instruction bits involved in each RT is comparatively high. Therefore, *version selection during compaction* is of high importance for DSPs. In [DLSM81], *version shuffling* was proposed as a technique for version selection, which can be integrated into heuristic algorithms: Whenever some RT x_i is to be assigned to a control step t, the cross product of all versions of x_i and all versions of RTs already assigned to t are checked for a combination of compatible versions. However, version shuffling does not permit to remove an "obstructive" RT from a control step t, once it has been bound to t, and therefore has a limited optimization effect.

Side effects: A side effect is an undesired RT, which may cause incorrect behavior of a machine program. Most compaction approaches assume, that the instruction format is such that side effects are excluded in advance. However, if arbitrary instruction formats are to be handled, three kinds of side effects must be considered during code compaction:

If the processor model comprises explicit tristate busses, **bus conflicts** may occur, whenever unused bus drivers are not deactivated for each control step. In this case, program behavior becomes unpredictable. In our approach, control code requirements for bus drivers are already determined during ISE, so that these are encoded in the RT versions before compaction starts. In this way, bus conflicts are mapped to usual encoding conflicts.

---

[8] A special case of resource conflicts may arise from mode register requirements: Suppose, that two RTs x_i, x_j are independent of each other and compatible w.r.t. data-path and instruction format, but require different states $m_i \neq m_j$ of a mode register R. Then, parallalization of x_i, x_j is obviously incorrect, but is not inhibited by partial instructions. Still, the dependency relations preserve semantical correctness during compaction: x_i is data-dependent on some RT x_k generating mode m_i, and analogously for x_j and some RT x_l. Let $k < l$, so that $x_k \xrightarrow{OD} x_l$. Then the relationship

$$x_k \xrightarrow{DD} x_i \xrightarrow{DAD} x_l \xrightarrow{DD} x_j$$

prevents x_i, x_j from being scheduled in the same control step, and analogously for $l < k$. Thus, if RTs for mode register setting are generated *before* compaction, incompatibility of partial instructions is a necessary and sufficient condition for inter-RT conflicts even in presence of mode registers.

Horizontal side effects occur in instruction formats, where a number of instruction bits remain don't care for each control step t. Whenever such a don't care bit steers a register or memory, which may contain a live value in step t, the register must be explicitly deactivated.

The third type of side effects, which we call **vertical side effects**, occurs in presence of encoded instruction formats. A vertical side effect is exposed, if a version b_i for an RT x_i is "covered" by a version b_j for another RT x_j, i.e. selection of b_i for x_i implies that x_j will be executed in the same control step. If x_j happens to be an RT ready to be scheduled, this side effect can be exploited. On the other hand, version selection must discard version b_i, whenever this is not the case, and x_j might destroy a live value. An example is the multiply-accumulate (MPYA) instruction on the TMS320C25 DSP (cf. section 1.4.2). Since there are also instructions that perform multiply and accumulate separately (MPY, APAC), selecting the MPYA version for either of the two implies a side effect. As shown in the DSPStone project [ZVSM94], even commercial compilers poorly exploit this type of parallelism. In our compaction technique, we avoid undesired side effects by carefully packing NOPs into each control step.

Time constraints: Since DSP algorithms are often subject to real-time constraints, compaction techniques are desirable, which parallelize RTs with respect to a given (maximum) time constraint of $T_{\max}$ machine cycles instead of aiming at optimal schedules. It might be the case, that a locally suboptimal scheduling decision leads to satisfaction of $T_{\max}$, while a rigid optimization heuristic fails. Time constraints could also be incorporated into preceding code generation phases, but should be taken into account *at least* during compaction, since only compaction makes the execution speed of a machine program exactly known.

Most of the related approaches to DSP code generation have adopted heuristic compaction algorithms [Wess96, LMP94a, LVKS+95, SuMa95] and thus show restrictions in the instruction formats that can be handled. In addition, no provisions are made for consideration of time constraints. Two exact (non-heuristic) compaction methods have been reported, which do take time constraints into account. Wilson's Integer Programming approach [WGHB94] (cf. section 4.3) integrates all low-level code generation phases including code compaction. The IP model comprises alternative versions and vertical side effects, but no bus conflicts and horizontal side effects. Furthermore, the IP model – at least in its entirety – turned out to be too complex for realistic target architectures, and the issue of retargetability was not treated. Timmer's graph-based NEAT scheduler [TSMJ95] achieves low runtimes for exact code compaction by

Instruction-level parallelism

pruning the search space in advance. NEAT produced very good results for a family of real-life ASIPs [SMT+95], but has restricted capabilities with respect to alternative versions and side effects, because it is based on the assumption, that all inter-RT conflicts are fixed before compaction. In presence of alternative versions, this means that incompatibility of versions b_i for RT x_i and b_j for RT x_j implies pairwise incompatibility of all versions for x_i and x_j. This assumption is often not fulfilled, and for instance may prevent exploitation of parallel AGU operations.

5.5.3 An IP model for DSP code compaction

In the following, we present a novel approach to exact local code compaction under time constraints, which meets the above-mentioned requirements. First, we give a modified definition of the code compaction problem, which captures alternative versions and side effects.

5.5.3.1 Definition

Let $BB = (x_1, \ldots, x_n)$ be an RTL basic block, where each x_i has a set $B_i = \{b_{i1}, \ldots, b_{in_i}\}$ of alternative versions (binary encodings). Furthermore, let $NOP = \{NOP_1, \ldots, NOP_r\}$ denote the set of no-operations for all registers $\{R_1, \ldots, R_r\}$ which are destinations of RTs in BB, and let $\{nop_{j1}, \ldots, nop_{jn_j}\}$ be the set of alternative versions for all $NOP_j \in NOP$.

A **(parallel) schedule** for BB is a sequence $CS = (CS_1, \ldots, CS_n)$, so that for any x_i, x_j in BB the following conditions hold:

- Each $CS_t \in CS$ is a subset of

$$\bigcup_{j=1}^{n} B_j \;\cup\; \bigcup_{j=1}^{r} NOP_j$$

- There exists exactly one $CS_t \in CS$, which contains a version of x_i (notation: $cs(x_i) = t$).

- If $x_i \xrightarrow{DD} x_j$ or $x_i \xrightarrow{OD} x_j$ then $cs(x_i) < cs(x_j)$.

- If $x_i \xrightarrow{DAD} x_j$, then $cs(x_i) \leq cs(x_j)$.

- If $x_i \xrightarrow{DD} x_j$, then all control steps $CS_t \in \{CS_{cs(i)+1}, \ldots, CS_{cs(j)-1}\}$ contain a NOP version for the destination of x_i.

- For any two versions

$$u, v \in \bigcup_{j=1}^{n} B_j \cup \bigcup_{j=1}^{r} NOP_j$$

there is no control step $CS_t \in CS$, for which

$$u, v \in CS_t \quad \wedge \quad u \not\sim v$$

For an RTL basic block BB whose RDG has critical path length L_c, **time-constrained code compaction** (TCC) is the problem of computing a schedule CS, such that, for a given $T_{\max} \in \{L_c, \ldots n\}$, CS satisfies $CS_t = \emptyset$ for all $t \in \{T_{\max}+1, \ldots, n\}$.

TCC is essentially the decision variant of *optimal* code compaction, extended by alternative versions and side effects, and is therefore NP-complete. Recently, several approaches have been published, which map NP-hard VLSI-design related problems into an Integer (Linear) Programming (IP) formulation (e.g. [GeEl92, LMD94, NiMa96]). IP [GaJo79] is the problem of computing a setting of n integer *solution variables* $(z_1, \ldots, z_n)$, such that a linear *objective function* $f(z_1, \ldots, z_n)$ is maximized (minimized) under the linear constraints

$$A \cdot (z_1, \ldots, z_n)^T \leq (\geq) B$$

for an $M \times N$ matrix A and an N-dimensional vector B. Although IP is NP-hard (the decision problem is NP-complete [GaJo79]), modelling intractable problems as Integer Programs can be a promising approach, because of the following reasons:

- Since IP is based on a relatively simple mathematical notation, its is easily verified, that the IP formulation of some problem meets the problem specification. Thus, the solutions are "correct by construction".

- IP is a suitable method for describing problems with many different types of constraints, because these constraints often have a straightforward representation in the form of linear inequations. Solving the IP means, that all constraints are *simultaneously* taken into account.

- Since IP is among the most important optimization problems, commercial tools are available for IP solving. Such IP solvers often compute solutions very fast, even for relatively large Integer Programs. Using an appropriate IP formulation thus often permits to optimally solve NP-hard problems of practical relevance.

In the following, we therefore develop an IP-based solution method for TCC. Assignment of RTs to control steps under given constraints as well as version selection for RTs is accomplished by solving an IP model of the compaction problem. In contrast to Wilson's approach [WGHB94], the IP is not created manually, but is automatically derived from the given problem instance and an externally specified time constraint. Furthermore, it focusses *only* on the problem of code compaction, which permits to compact machine code of comparatively complex DSPs within acceptable amounts of computation time.

The solution variables in the IP model encode the RT versions contained in each control step, and the constraints ensure correctness of the schedule according to definition 5.5.3.1. Solution variables are only defined for control steps less or equal $T_{\max}$, so that there is no objective function, but just constraint satisfaction is required. The setting of solution variables also accounts for NOPs, that have to be activated for each control step. Necessary NOPs are inserted into control steps by means of a version shuffling mechanism.

Given an instance of TCC, first the *mobility range*

$$rng(x_i) = [ASAP(x_i), ALAP(x_i)]$$

is determined for each RT x_i. In contrast to definition 5.5.1.2, we use $T_{\max}$ as the maximum ALAP time for all x_i, instead of the critical path length L_c. The solution variables are subdivided into two classes of indexed *decision* (0/1) variables:

V-variables (triple-indexed): For each x_i with version set B_i the following *V-variables* are defined:

$$\{v_{i,m,t} \mid m \in \{1,\ldots,|B_i|\} \land t \in rng(x_i)\}$$

The *interpretation* of V-variables is

$$v_{i,m,t} = 1 \quad :\Leftrightarrow$$

RT x_i is scheduled in control step number t with version $b_{im} \in B_i$

N-variables (double-indexed): For the set $NOP = \{NOP_1, \ldots, NOP_r\}$ of no-operations, the following *N-variables* are defined:

$$\{n_{s,t} \mid s \in \{1,\ldots,r\} \land t \in [1, T_{\max}]\}$$

The *interpretation* of N-variables is

$$n_{s,t} = 1 \quad :\Leftrightarrow$$

Control step number t contains a NOP for destination register R_s

The correctness conditions are encoded into IP constraints as follows:

Each RT is scheduled exactly once: This is ensured, if the sum over all V-variables for each RT x_i equals 1.

$$\forall x_i : \sum_{t \in rng(x_i)} \sum_{m=1}^{|B_i|} v_{i,m,t} = 1$$

Data- and output-dependencies are not violated: If x_i, x_j are data- or output-dependent, and x_j is scheduled in control step t, then x_i must be scheduled in an earlier control step, i.e. in the interval $[ASAP(x_i), t-1]$.

$$x_i \xrightarrow{DD} x_j \quad \vee \quad x_i \xrightarrow{OD} x_j \quad \Rightarrow$$

$$\forall t \in rng(x_i) \cap rng(x_j) : \sum_{m=1}^{|B_j|} v_{j,m,t} \leq \sum_{t' \in [ASAP(x_i), t-1]} \sum_{m=1}^{|B_i|} v_{i,m,t'}$$

Data-anti-dependencies are not violated: This is similar to the previous case, except that x_i may also be scheduled in parallel to x_j.

$$x_i \xrightarrow{DAD} x_j \quad \Rightarrow$$

$$\forall t \in rng(x_i) \cap rng(x_j) : \sum_{m=1}^{|B_j|} v_{j,m,t} \leq \sum_{t' \in [ASAP(x_i), t]} \sum_{m=1}^{|B_i|} v_{i,m,t'}$$

Live values must not be destroyed by side effects: A value in a single register R_s is live in all control steps between its production and consumption, so that a NOP must be activated for R_s in these control steps. As in the MSSQ compiler, we assume that a NOP version for R_s can be selected *after* compaction for each control step, in which R_s is not written, i.e. a compatible NOP version can always be found by means of version shuffling. Although one can construct counterexamples, this assumption is fulfilled for real-life instruction formats. We permit to *tolerate* side effects, i.e. NOPs for registers are activated only if two data-dependent RTs are not scheduled in consecutive control steps. Conversely, we *enforce* to schedule these RTs consecutively, if no NOP for the corresponding destination register exists. This is modelled by the following constraints:

$$R_s = destination(x_i) \quad \wedge \quad x_i \xrightarrow{DD} x_j \quad \Rightarrow$$

$$\forall t \in \underbrace{[ASAP(x_i)+1, ALAP(x_j)-1]}_{=:R(i,j)}:$$

$$\sum_{t' \in R(i,j) | t' < t} \sum_{m=1}^{|B_i|} v_{i,m,t'} + \sum_{t' \in R(i,j) | t' > t} \sum_{m=1}^{|B_j|} v_{j,m,t'} - 1 \leq n_{s,t}$$

The left hand side of the inequation becomes 1, exactly if x_i is scheduled before t, and x_j is scheduled after t. In this case, a NOP version for R_s must be activated in t. If no NOP is present for register R_s, then $n_{s,t}$ is replaced by zero. This mechanism is only useful for *single* registers. Tolerating side effects for (addressable) register *files* is only possible, if N-variables are introduced for *each element* of the file, because the different elements must be distinguished. However, this would imply an intolerable explosion of the number of IP solution variables. Instead, we assume that a NOP is present for each register file, which in practice is no severe restriction.

Compatibility restrictions are not violated: Two RTs x_i, x_j have a potential conflict, if they have at least one pair of conflicting versions, they have non-disjoint mobility ranges, and they are neither data- nor output-dependent. The following constraints ensure, that at most one of two conflicting versions is scheduled in each control step t.

$$\forall x_i, x_j, \quad (x_i, x_j) \notin DD \cup OD: \quad \forall t \in rng(x_i) \cap rng(x_j):$$

$$\forall b_{im} \in B_i: \quad \forall b_{jm'} \in B_j: \quad b_{im} \not\sim b_{jm'} \Rightarrow v_{i,m,t} + v_{j,m',t} \leq 1$$

A basic block BB has a parallel schedule with $T_{\max}$ control steps, exactly if for the corresponding IP there exists a 0/1 setting of V and N-variables, so that all constraints are satisfied. In this case, the actual schedule can be directly derived from the set of solution variables that are set to 1. It is important to keep the number of solution variables as low as possible in order to reduce the runtime requirements for IP solving. The number of variables can be reduced by discarding redundant variables in advance: Obviously, all N-variables which do not appear in "live value" constraints are redundant. Furthermore, alternative versions for RTs are only useful, if they potentially increase parallelism. Scheduling version b_{im} of some RT x_i in control step t potentially increases parallelism, whenever there exists an RT x_j, which could be scheduled in parallel, i.e. x_j meets the following conditions:

1. $t \in rng(x_j) \quad \wedge \quad (x_i, x_j) \notin DD \cup OD$, and
2. x_j has a version $b_{jm'}$ compatible to b_{im}

If no such x_j exists, then all variables $v_{i,m,t}$ are (for all m) equivalent in terms of parallelism, and it is sufficient to keep only one arbitrary representative (e.g. $v_{i,1,t}$).

A further reduction of the number of variables can be achieved through computing tighter mobility ranges by applying some ad hoc rules. For instance, if $x_i \xrightarrow{DAD} x_j$, and all versions for x_i, x_j are pairwise incompatible, then $ASAP(x_j)$ has a lower bound of $ASAP(x_i) + 1$.

The setting of IP solution variables accounts for the control step assignment of RTs. Insertion of NOPs is performed as described in fig. 5.16. Algorithm

```
(1)    algorithm INSERTNOPS
(2)    input: schedule CS = {CS_1, ..., CS_{T_max}}
(3)    output: schedule CS, augmented with NOPs
(4)    begin
(5)        for t = 1 to T_max do
(6)            for all single registers R_s do
(7)                if n_{s,t} = 1 then SHUFFLEVERSIONS(NOP_s, CS_t);
(8)            end for
(9)            for all addressable registers R_a do
(10)               X(R_a) := {x_i | destination(x_i) = R_a};
(11)               if ∑_{x_i ∈ X(R_a)} ∑_{m=1}^{|B_i|} v_{i,m,t} = 0
(12)               then SHUFFLEVERSIONS(NOP_a, CS_t);
(13)           end for
(14)       end for
(15)       return CS;
(16)   end algorithm
```

Figure 5.16 Insertion of NOPs into parallel RT schedules

INSERTNOPS processes all control steps t between 1 and $T_{\max}$ separately. If, for some single register R_s, $n_{s,t} = 1$ indicates that a NOP has to be activated in control step CS_t, then an appropriate NOP version is added to CS_t by subroutine SHUFFLEVERSIONS. SHUFFLEVERSIONS examines the cross-product of all versions of RTs assigned to CS_t and all NOP versions for R_s in order to find a compatible combination of versions. Existence of this combination is guaranteed by assumption. For addressable registers R_a, the set $X(R_a)$ of all

Instruction-level parallelism

RTs writing to R_a is determined. If no RT in $X(R_a)$ is scheduled in step CS_t, as indicated by the setting of V-variables, then a NOP for R_a is added to CS_t in order to prevent side effects. This is accomplished again by procedure SHUFFLEVERSIONS.

Since all versions are associated with a binary partial instruction, the final result of code compaction and NOP insertion is a binary machine code listing, which can be regarded as a program memory initialization. In order to obtain executable machine code, all control steps have to be completed by partial instructions for program counter modification. If a control step CS_t comprises a branch operation, then the binary jump address has to be added, whose value is known after compaction. For control steps not containing a branch, a version has to be added, which represents an increment operation on the program counter. As for NOPs, it is reasonable to assume that these versions can be packed *after* compaction, which can be easily integrated into algorithm INSERTNOPS. This postprocessing does not require special techniques, and is thus not further detailed here.

5.5.4 Example

A potential weakness of exact IP-based code compaction is the amount of computation time needed for IP solving. Therefore, this section presents experimental results in order to indicate the problem complexities (in terms of basic block length and target machine complexity) which can be handled by our approach. We demonstrate exploitation of parallelism for the TI 'C25 target processor. We consider code generation for the complex_multiply program from the DSPStone benchmark set [ZVSM94], while also focussing on the cooperation between compaction and address assignment. The source program computes the product c of two complex numbers a, b and is given by two lines of code:

```
cr = ar * br - ai * bi ;
ci = ar * bi + ai * br ;
```

Column 1 in fig. 5.17 shows the vertical machine code for complex_multiply, as generated by tree parsing and scheduling according to the techniques presented in the previous chapter. The code sequentially computes the real and imaginary parts of the result and employs the TMS320C25 registers TR, PR, and ACCU. The variables are not yet bound to memory addresses. The variable access sequence in the vertical code is (ar, br, ai, bi, cr, ar, bi, ai, br, ci). For a

```
TR = MEM[ar]         (1)    ARP = 0                    // init AR pointer
PR = TR * MEM[br]    (2)    AR[0] = 5                  // AR = 5 (ar)
ACCU = PR            (3)    TR = MEM[AR[ARP]]          // TR = ar
TR = MEM[ai]         (4)    AR[ARP] -= 4               // AR = 1 (br)
PR = TR * MEM[bi]    (5)    PR = TR * MEM[AR[ARP]]     // PR = ar * br
ACCU = ACCU - PR     (6)    ACCU = PR                  // ACCU = ar * br
MEM[cr] = ACCU       (7)    AR[ARP] ++                 // AR = 2 (ai)
TR = MEM[ar]         (8)    TR = MEM[AR[ARP]]          // TR = ai
PR = TR * MEM[bi]    (9)    AR[ARP] ++                 // AR = 3 (bi)
ACCU = PR            (10)   PR = TR * MEM[AR[ARP]]     // PR = ai * bi
TR = MEM[ai]         (11)   ACCU = ACCU - PR           // ACCU = ar * br - ai * bi
PR = TR * MEM[br]    (12)   AR[ARP] ++                 // AR = 4 (cr)
ACCU = ACCU + PR     (13)   MEM[AR[ARP]] = ACCU        // cr = ar * br - ai * bi
MEM[ci] = ACCU       (14)   AR[ARP] ++                 // AR = 5 (ar)
                     (15)   TR = MEM[AR[ARP]]          // TR = ar
                     (16)   AR[ARP] -= 2               // AR = 3 (bi)
                     (17)   PR = TR * MEM[AR[ARP]]     // PR = ar * bi
                     (18)   ACCU = PR                  // ACCU = ar * bi
                     (19)   AR[ARP] --                 // AR = 2 (ai)
                     (20)   TR = MEM[AR[ARP]]          // TR = ai
                     (21)   AR[ARP] --                 // AR = 1 (br)
                     (22)   PR = TR * MEM[AR[ARP]]     // PR = ai * br
                     (23)   ACCU = ACCU + PR           // ACCU = ar * bi + ai * br
                     (24)   AR[ARP] --                 // AR = 0 (ci)
                     (25)   MEM[AR[ARP]] = ACCU        // ci = ar * bi + ai * br
```

Figure 5.17 Vertical code for complex_multiply and TMS320C25 machine instructions: left/center column: before and after insertion of AGU operations, right column: register contents.

single address register, the general offset assignment algorithm from section 5.3.2 computes the permutation

$$0 \leftrightarrow ci, \quad 1 \leftrightarrow br, \quad 2 \leftrightarrow ai, \quad 3 \leftrightarrow bi, \quad 4 \leftrightarrow cr, \quad 5 \leftrightarrow ar$$

Column 2 in fig. 5.17 shows the vertical code, augmented by the set of AGU operations resulting from the chosen address assignment. The comments in the third column indicate the register contents after each RT. The total number of RTs including AGU operations is 25, with a critical path length of 15. Up to 30 different versions are available for the the RT patterns occurring in the vertical code, so that a valid parallel schedule is not obvious.

$T_{\max}$	CPU sec	solution	# V-variables	# N-variables
15	8.39	no	71	23
16	0.75	yes	141	44
17	1.26	yes	211	56
18	22	yes	281	64
19	119	yes	351	72
20	164	yes	421	79
21	417	yes	491	84

Table 5.4 Experimental results for IP-based compaction of complex_multiply/TMS320C25 code

Table 5.4 shows experimental data (CPU seconds[9], number of V- and N-variables) for IP-based compaction of the complex_multiply code for $T_{\max}$-values in [15, 21]. For the theoretical lower bound $T_{\max} = 15$, induced by the critical path length, no solution exists, while for $T_{\max} = 16$ (the actual lower bound) a schedule is constructed in less that 1 CPU second. Beyond $T_{\max} = 18$, the CPU time rises to minutes, due to the large search space that has to be investigated by the IP solver. This unfavorable effect is inherent to any IP-based formulation of a time-constrained scheduling problem: The computation time may dramatically grow with the number of control steps, even though the scheduling problem intuitively gets easier. Therefore, it is favorable to choose relatively tight time constraints, i.e. close to the actual lower bound on the schedule length. In this case, IP-based compaction produces very high quality code within acceptable amounts of computation time: Fig. 5.18 shows the (optimal) parallel schedule constructed for $T_{\max} = 16$. The code quality is the same as achieved by the Texas Instruments C compiler: Although the TI compiler does not exploit instruction-level parallelism, utilization of direct addressing capabilities saves address register initialization and immediate modify operations. The hand-crafted reference assembly code for complex_multiply comprises 19 instructions. The reference code does exploit potential parallelism (e.g. the "LTP" instruction, cf. fig. 5.18), but usage of too many address registers causes overhead. In turn, this stresses the importance of address assignment as a preparation step before code compaction.

---

[9] Integer Programs have been solved with *Optimization Subroutine Library* (OSL) V1.2 on a SPARC-20. The CPU times for IP file generation and insertion of NOPs are negligible and thus not mentioned.

```
(1)         ARP = 0                                           LARP 0
(2)         AR[0] = 5                                         LARK AR0,5
(3)         TR = MEM[AR[ARP]]                                 LT *
(4)         AR[ARP] -= 4                                      SBRK 4
(5,7)       PR = TR * MEM[AR[ARP]] || AR[ARP] ++              MPY *+
(6,8,9)     ACCU = PR || TR = MEM[AR[ARP]] || AR[ARP] ++      LTP *+
(10,12)     PR = TR * MEM[AR[ARP]] || AR[ARP] ++              MPY *+
(11)        ACCU = ACCU - PR                                  SPAC
(13,14)     MEM[AR[ARP]] = ACCU || AR[ARP] ++                 SACL *+
(15)        TR = MEM[AR[ARP]]                                 LT *
(16)        AR[ARP] -= 2                                      SBRK 2
(17,19)     PR = TR * MEM[AR[ARP]] || AR[ARP] --              MPY *-
(18,20,21)  ACCU = PR || TR = MEM[AR[ARP]] || AR[ARP] --      LTP *-
(22,24)     PR = TR * MEM[AR[ARP]] || AR[ARP] --              MPY *-
(23)        ACCU = ACCU + PR                                  APAC
(25)        MEM[AR[ARP]] = ACCU                               SACL *
```

Figure 5.18 Optimal parallel TMS320C25 code for complex_multiply: The right column shows TMS320C25 assembly syntax ("*" denotes memory access by the current address register, "*+, *-" denote auto-increment/decrement).

5.5.5 Optimal local compaction

The concept of considering code compaction as a time-constrained problem is useful, whenever the source algorithm consists of a single block, and a user-specified constraint on the maximum number of instructions is at hand. While this scenario is present for many DSP applications, a modification is necessary for large basic block lengths, which cannot be handled by IP-based compaction at once. If an acceptable amount of computation time were exceeded, then it is necessary to split basic blocks into sequences of subblocks, which are compacted separately. The (possibly suboptimal) global solution is then obtained by concatenating compacted subblocks. The maximum acceptable block size – which depends on the particular target architecture and instruction format – can be passed as a parameter to the code compaction procedure.

Presence of subblocks poses the additional problem of appropriately distributing the "machine cycle budget" $T_{\max}$ over different subblocks, so that the global constraint is met. The same problem is encountered for programs, that consist of multiple basic blocks. In such a situation, it can hardly be determined in advance, how a fixed number of machine cycles must be distributed over all blocks. In general, the actual critical path through all control-flow paths of a program is not known before compaction, because the block-local lower bounds reflected by critical path lengths in RT dependency graphs can usually not be

Instruction-level parallelism

reached due to encoding conflicts. A pragmatic approach to solve this problem is usage of locally optimal code compaction procedure, which – regardless of a time constraint – constructs optimal parallel schedules for RTL basic blocks. Then, the worst-case execution time through the different paths of control flow in a multi-block program can be verified against a global time constraint.

The presented IP model for time-constrained code compaction can also be used for locally optimal code compaction. We briefly discuss three different options.

Linear search over schedule lengths

The length of any valid parallel schedule must be in between the critical path length and the total number of RTs. Therefore, one can iterate time-constrained compaction over all $T \in [L_c, n]$. Since tight time constraints in general lead to smaller solution times, the iteration should start at L_c, and the iteration can be terminated as soon as the first solution was detected, which yields the optimal schedule. In practice, linear search is a good choice in terms of computation time, because feasibility for small $T_{\max}$ values is usually checked quickly.

Binary search over schedule lengths

Alternatively, one can perform a binary search over schedule lengths in the interval $[L_c, n]$. Whether or not binary search outperforms linear search in terms of computation time depends on the difference between the theoretical lower bound L_c and the actual lower bound. If this difference is small, then binary search investigates several solutions requiring high computation times, before the optimum is detected. In contrast, binary search is better for machines with very restricted parallelism, for which the optimum schedule length does not significantly differ from n.

Minimization

The IP formulation of compaction as a decision problem can also be transformed into an optimization problem by some minor extensions of the IP model. A dummy RT x_{n+1} with a single version $b_{n+1,1}$ is introduced, and is marked as data-dependent on all RTs $x_1, \ldots, x_n$, so that x_{n+1} is enforced to be assigned to the last control step. The mobility range of x_{n+1} is $[L_c + 1, n + 1]$. The constraint

$$\sum_{t=L_c+1}^{n+1} v_{n+1,1,t} = 1$$

ensures that x_{n+1} is assigned to one of these control steps. Then, the objective function

$$f \;=\; 1 * v_{n+1,1,L_c+1} \;+\; \ldots \;+\; (n - L_c + 1) * v_{n+1,1,n+1}$$

is minimized exactly if the RTs $x_1, \ldots, x_n$ are optimally compacted. This optimization variant has the advantage, that the complete search procedure is left to the IP solver. Unfortunately, in practice this may be over-compensated by the fact, that the IP model must contain solution variables for the complete interval $[1, n]$ of control steps. Therefore, "real" minimization is often excluded due to prohibitively high runtimes. Compacting the complex_multiply code from fig. 5.17 by means of direct minimization, for instance, requires 137 CPU seconds, while the linear search strategy detects the optimum of 16 instructions after $8.39 + 0.75 = 9.14$ seconds (cf. table 5.4).

The RECORD compiler, which is presented in the next chapter, offers different strategies for IP-based code compaction. For single-block programs, the user may either specify a fixed maximum time constraint or select locally optimal compaction. For multi-block programs, only the latter mode is permitted, for which all three optimization methods described above are available. Which of these modes is most appropriate depends both on the target processor and the application. Therefore, the selection of compaction strategies is left to the user.

6
THE RECORD COMPILER

The code generation techniques presented in the previous chapters have been implemented in form of the prototype compiler system RECORD. The current version is implemented in C++ under a UNIX workstation environment. It consists of a package of separate programs, which are invoked from a graphical user interface, and which communicate via file exchange formats. The total amount of C++ source code is approximately 120,000 lines. Out of these, about 30 % are spent for the necessary language frontends, 50 % for code generation, and 20 % for instruction set extraction including tree parser generation. In this chapter we exemplify, how RECORD's concept of retargetability can be exploited in order to customize processor hardware towards a certain application. Afterwards, the code quality achieved with RECORD is evaluated for a standard DSP and a set of benchmark programs.

6.1 RETARGETABILITY

A major argument for using flexible, retargetable compilers is the fact, that – in contrast to off-the-shelf DSPs – application-specific programmable components may still be customized by the design engineer, so as to achieve a processor architecture optimized for a particular DSP application. Due to external constraints (e.g. enforced reuse of layout cells, restrictions of silicon technology or simulation tools) the coarse architecture might be fixed, but the designer still may influence the number and interconnect of functional units or registers. Such a design scenario is, for instance, present for audio signal processing ASIPs used at Philips [SMT+95].

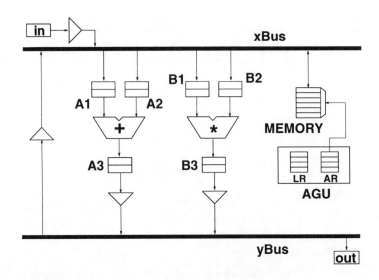

Figure 6.1 Coarse architecture of lattice filter ASIP

By re-compiling the application source code onto different detailed target architectures, the user of a retargetable compiler is capable of exploring a number of different design options within the possibly huge design space. Typically, under given performance constraints, one would like to minimize the required hardware area. In this section, we exemplify how the retargeting concept in RECORD supports such a design scenario. The example is very simple, and we only consider a small part of the actual design space. For sake of simplicity, here we neglect the impact of architectural changes on the clock period length. What we want to demonstrate is, that design space exploration can be performed quickly, only based on the designer's "intuition", instead of a deep analysis of source code and target architecture characteristics.

Suppose, that an ASIP needs to be designed for the lattice filter mentioned in chapter 4 (fig. 4.4). The coarse target architecture specification is illustrated in fig. 6.1. It contains functional units connected by tristate busses. Functional units are optionally equipped with register files at the inputs and outputs. We assume a horizontal microprogrammable controller with single-cycle instructions. Addresses for the on-chip memory are generated by an AGU with a flexible amount of address and length registers. Ring buffers can be simulated by modulo addressing.

The RECORD compiler 181

Suppose that, due to area limitations, the number of functional units (here: one adder, one multiplier) must not be exceeded, and that a real-time constraint demands, that for each sample period the filter output has to be available in the output register after a maximum of 30 machine cycles. For these constraints, we examine six different detailed configurations of the data-path architecture, which are depicted in fig. 6.2. The processor model for each configuration contains approximately 250 lines of MIMOLA code. Since a processor schematic is available, we use an RT-level processor model based on the specification in fig. 6.1.

Configuration 1: In order to get a first idea of whether the performance constraint can be met, we compile the lattice filter source to a configuration, where all input/output register files of functional units have a size of one. The necessity of an adder and a multiplier is obvious from the source code. Since there are two delay lines in the filter, the AGU must contain exactly two buffer length registers. We start with four address registers, two of which are required for modulo addressing, and the remaining two are intended for addressing other scalar values. Compilation by RECORD yields a vertical code length of 65 RTs, which cannot be compacted into 30 control steps. Further compaction runs show, that 39 control steps would be the minimum for this configuration. The report file generated by the address generator, however, shows that the overhead is not due to the AGU configuration, because a zero-cost address assignment for scalar values is achieved even with a single address register, except for those two needed for ring buffers. Therefore, we can fix the AGU configuration and focus on the pure data-path.

Configuration 2: The source code essentially consists of a set of sum-of-product computations. Therefore, it might be a good idea, if products computed by the multiplier could be directly passed to the adder, without taking a detour over registers and busses. In configuration 2, we therefore insert a direct connection between the multiplier and adder modules. As a consequence, the code generator may decide to execute multiply-accumulates in chained mode. Compilation of the source code to this configuration yields vertical code with 67 RTs, and the performance constraint is still not met. However, the best parallel schedule would consist of only 37 instructions in this case, which shows that chaining would be a better solution.

Configuration 3: Next, we explore a more unconventional configuration. Instead of the common multiply-accumulate operation, we try an add-with-multiply chain. That is, a direct connection is introduced between the adder and the multiplier. This could be reasonable, because both summation of products and multiplication of sums are present in the source code. The resulting

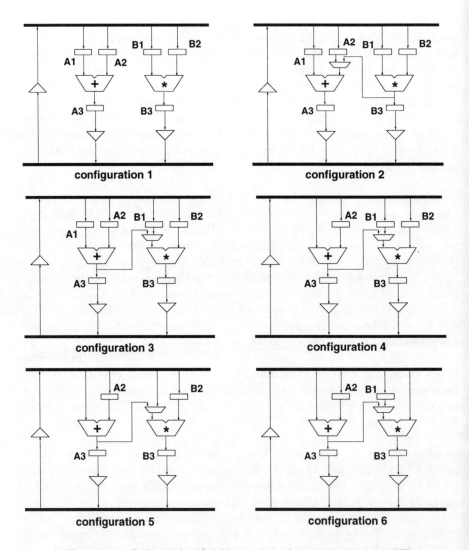

Figure 6.2 Different detailed data-path configurations of target ASIP

code brings us closer to satisfying the performance constraint. The vertical code of 62 RTs can be compacted to 35 control steps. Therefore, we prefer to keep the add-with-multiply chain instead of configuration 2.

Configuration 4: Obviously, a performance penalty could also be due to superfluous data transfer operations in the machine code. Still, some values

configuration	vertical code length	parallel code length	CPU seconds (SPARC-20)
1	65	39	45.3
2	67	37	4.6
3	62	35	20.0
4	53	30	19.4
5	52	33	14.1
6	48	30	4.5

Table 6.1   Experimental results of ASIP design study

have to pass the input registers of functional units. Also the area overhead introduced by chaining in configuration 4 could enforce to remove hardware components. Therefore, we drop the input register A1 of the adder and replace it by a direct connection to the xBus. Compilation shows, that this is a favorable decision. The resulting machine code consists of only 53 RTs, which are compacted into 30 control steps. Therefore, configuration 4 is the first that meets the performance constraint.

Configuration 5: Next, we have additionally removed the input register B1 of the multiplier, in order to determine whether this would still result in a feasible solution at reduced area costs. The vertical code for this configuration consists only of 52 RTs, but the time constraint is no longer met. The minimum number of control steps generated by compaction is 33. Obviously, additional data dependencies, that originate from the decision to remove B1, lead to an increased machine cycle count.

Configuration 6: In the last configuration, the input register B2 for the right multiplier input has been removed instead, and has been replaced by a direct bus connection. In contrast to configuration 5, this detailed target architecture again meets the performance constraint. Compilation by RECORD yields 48 RTs in the vertical code, which are compacted into 30 control steps. Compared to the original configuration 1, we have saved 2 registers and 9 control steps, at the expense of an additional multiplexer.

The experimental data, including compilation times, are summarized in table 6.1. Though this is merely a "toy" example, it was shown how parts of the design space can be explored with RECORD, based on some coarse knowledge about the mutual dependence between data-path structures and code length.

The point is, that this exploration can be performed within very short time. Switching from one configuration to another requires only marginal local adaptations of the processor model. The automated retargeting procedure implemented in RECORD permits to evaluate a single configuration within a few minutes. Processor model analysis, instruction-set extraction, and tree parser generation take less than 5 SPARC-20 CPU seconds for this example. For no configuration, the compilation time (most of which is spent during compaction) exceeds 1 CPU minute.

6.2 CODE QUALITY

6.2.1 Experimentation methodology

The code quality achieved by a compiler can only be assessed, if a set of reference points is available, to which the results can be compared. Furthermore, in the context of retargetability, the processor architecture, for which code is generated, needs to be specified, because the code quality may depend on the target processor itself. Therefore, we evaluate the quality of RECORD-generated code for a well-documented standard DSP, namely the Texas Instruments TMS320C25 [TI90], which was already mentioned several times in the previous chapters.

Evaluation of code quality achieved by commercial C compilers for a set of standard DSPs has been the goal in the DSPStone project [ZVSM94]. The DSP-Stone methodology is based on comparison of compiler-generated code against hand-crafted reference code. The DSPStone benchmark programs are a collection of relatively small C programs, which were selected as representatives of pieces of arithmetic-intensive code, which are typical for DSP algorithms, such as multiply-accumulates on complex numbers, or convolution of vectors. Hand-written assembly code for larger programs is usually not publicly available. However, since larger DSP programs essentially are compositions of such small code pieces, the DSPStone results provide an estimation of the quality of compiler-generated code for realistic applications. Furthermore, the limited size of the benchmark programs facilitates analysis of machine programs, so as to identify strengths and weaknesses of different code generation techniques.

Based on these observations, we discuss the code quality achieved by RECORD for DSPStone benchmark programs, while using three reference points:

The RECORD compiler

1. the manually programmed assembly code

2. the code generated by the Texas Instruments TMS320C2x/C5x ANSI C Compiler V6.40 (highest optimization effort)

3. the code generated by the MSSQ compiler (section 2.2).

Before presenting the experimental data, it is necessary to focus on two problems, which potentially bias the obtained results.

The first problem arises from the use of **different source languages**: Some DSPStone C programs are written in low-level style. In particular, the programmers made heavy use of pointer arithmetic. The TI C compiler recognizes C expressions of type "*ptr++" and translates these into indirect addressing constructs with parallel auto-increment operations. Such pointer-oriented source codes were manually translated into array-oriented DFL programs, in order to permit compilation by RECORD. The array-oriented DFL code is much easier to read than pointer-oriented C, and the code quality must be interpreted while keeping in mind the effort spent in source code specification. Unfortunately, the DSPStone documentation provides no information on how using arrays instead of pointers affects code quality.

The target processor in DSPStone was the TI TMS320C51 DSP, which is a successor of the TMS320C25. In contrast to the 'C25, the TMS320C51 supports **zero-overhead loops**, which cover **multi-instruction loop bodies**. For the reasons explained in section 4.6.3, the current RECORD implementation does not exploit zero-overhead loops. However, all generated 'C51 instructions *not* related to loop control are also 'C25 instructions. The 'C25 only supports zero-overhead repetition of a single instruction. For multiple-instruction loop bodies, the C compiler thus needs to generate extra instructions for loop control, similar to the conditional branch scheme used in RECORD (cf. section 4.6.3). However, the DSPStone documentation does not indicate the amount of instructions saved by zero-overhead loops compared to an equivalent implementation by conditional branches. As a consequence, results concerning 'C25 code *including loop control overhead* are not available. Thus, for experimental evaluation, we neglect instructions generated for loop control (i.e. loop counter initialization and updates) but focus on instructions generated for useful computations. For the latter, an unbiased comparison is possible.

benchmark	TI	ref.	RECORD	CPU	MSSQ	CPU
real_update $D = C + A * B$ (real numbers)	6	10	6	3.7	9	56
complex_multiply $C = A * B$ (complex numbers)	16	19	15	10.2	26	68
complex_update $D = C + A * B$ (complex numbers)	31	21	18	8.6	36	76
N_real_updates $i = 1 \ldots N$: $D[i] = C[i] + A[i] * B[i]$ (real numbers)	9	5	5	119	–	–
N_complex_updates $i = 1 \ldots N$: $D[i] = C[i] + A[i] * B[i]$ (complex numbers)	31	17	20	91	–	–
fir (per tap) FIR filter (section 1.4.1)	7	1	2	96 (8 taps)	–	–
iir_biquad_one_section $w = x - a1 * w@1$ $\quad -a2 * w@2$ $y = b0 * w + b1 * w@1$ $\quad +b2 * w@2$	26	20	29	12.5	44	81
iir_biquad_N (N iir biquad sections)	42	12	31	49	–	–
dot_product $z = \vec{A} \cdot \vec{B}$ (two-dimensional vectors)	6	5	6	33	–	–
convolution $i = 1 \ldots N$: $y = y + x[i] * h[N - i]$	5	1	6	10.5	–	–

Table 6.2 Absolute results for DSPStone benchmark programs and TMS320C25 target processor

6.2.2 Experimental results

Experimental results for the DSPStone benchmark programs are listed in table 6.2. Column 1 gives the program name and a brief description of the program functionality. Columns 2 and 3 show the number of instructions generated by the TI C compiler and the length of the hand-written reference code, respectively. Columns 4 and 5 give the number of instructions generated by RECORD, as well as the required compilation time in SPARC-20 CPU seconds. For code compaction, the linear search strategy described in section 5.5.5 was applied, and IBM's OSL software was used for IP solving. Finally, columns 6 and 7 show the corresponding data obtained with the MSSQ compiler (cf. section 2.2)[1].

The achieved code quality can be assessed by means of the chart shown in fig. 6.3. The bars show the relative code size produced by the TI C Compiler, RECORD, and MSSQ, where the size of the hand-written reference code is set to 100 %. In some cases, the compiler-generated code is better than the reference code (i.e. < 100 %). These surprising results are encountered for benchmarks, for which instructions related to address generation are dominant: In the reference code, no attempt has been made to minimize the number of used address registers. As a result, initialization of address registers causes overhead, which can be avoided by address assignment techniques or – in case of the TI compiler – by direct addressing, which requires no initializations at all.

For most benchmarks, the size of code generated by RECORD does not exceed the size of hand-written code significantly. This is due to the fact, that the tasks of code selection, address generation, and code compaction, which are well solved in the manually programmed code, can be automated by the algorithms and techniques presented in the previous chapters. In the reference code, some instructions are saved by performing well-known simple global optimizations, which have not been treated in this book. An example is moving of loop-invariant computations outside of loop iterations.

For the benchmarks `fir`, `iir_biquad_N_sections`, and `convolution`, however, the overhead of compiler-generated code compared to hand-written code is very high (more than 100 %), in spite of the mentioned optimization phases.

In case of `fir` and `convolution`, the quality loss is caused by a single effect, namely a very DSP-specific instruction. The TMS320C2x/5x offers a "MAC" instruction intended for pipelined block processing, which executes up to 6 RTs in parallel. Exploitation of "MAC" in a compiler would demand for a *loop folding* technique capable of folding iterations *after* RTL code has been generated.

---

[1] For benchmarks comprising loop structures, no MSSQ results are available. This is due to restricted capabilities of MSSQ, which cannot exploit DSP-specific address generation units in case of high-level programs.

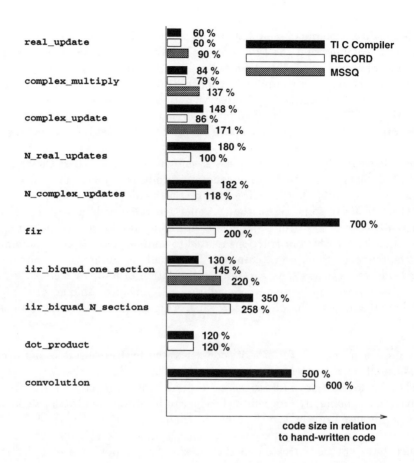

Figure 6.3 Relative results for DSPStone benchmark programs and TMS320C25 target processor

In case of the iir_biquad_N_sections program, the reasons for the prohibitively high compiler overhead are more complex. For the TI compiler, the unfavorable effects of missing data-flow analysis, inflexible address generation and ignoring potential parallelism accumulate for this benchmark. For RECORD, the overhead is lower, but still results from a phase coupling problem. The array accesses in the generated vertical code are not well arranged with respect to exploitation of parallel address register increments. The resulting extra instructions, which are not present in the hand-crafted code, contribute much to the observed overhead. This could presumably be avoided by closer coupling of RT scheduling and address generation.

6.2.3 Evaluation

The experimental results presented in this section indicate, that the code generation and optimization techniques presented in this book apply to realistic target machines. The RECORD compiler was successfully retargeted to a representative standard DSP. On the average, the achieved code quality is located between the hand-written reference code and code generated by a commercial, target-specific compiler.

This achievement comes at the price of increased compilation time, which is, however, not the most critical resource in DSP code generation. According to current estimations [Goos96], code quality overheads of 20–30 % of compiler-generated code versus hand-written code are acceptable under industrial conditions. For the benchmarks considered here, RECORD meets this demand in most cases. Inclusion of additional standard optimizations, such as keeping track of register contents across single expressions or loop iterations, would generate acceptable results also in many cases, where purely local techniques are insufficient. The analysis of the machine code generated by RECORD shows, that *local* code generation, comprising code selection, address generation, and code compaction for basic blocks, is well solved by the proposed techniques. Such local techniques could be embedded as subroutines into global techniques, operating beyond basic block boundaries, which are beyond the scope of this book. Nevertheless, highly DSP-specific machine instructions may still make it impossible to completely replace the human programmer by currently available compiler technology.

7
CONCLUSIONS

7.1 CONTRIBUTIONS OF THIS BOOK

Although the presented techniques have been implemented in form of a complete compilation procedure, which reads high-level source code and emits executable machine code, RECORD currently is not a complete compiler. For instance, the huge amount of known standard compiler optimizations have not been exploited, because new scientific insight would presumably not have resulted. The essential contribution of this book can be identified as a collection of *new and improved code generation techniques*. Nevertheless, the RECORD implementation proves, that these techniques could be immediately used as subroutines in realistic DSP compilers. The main achievements compared to previous work are summarized in the following.

7.1.1 Processor modelling and retargeting

Modelling of different realistic processors gives evidence of the necessity of *versatile* processor modelling capabilities. That is, abstraction levels ranging from the RT-level to the instruction/behavioral level need to be supported, in order to guarantee high modelling freedom for the user. Restricting the model abstraction level *implies restricting the class of target processors*, that can be handled in practice. A powerful processor modelling and model analysis mechanism is provided by the *instruction-set extraction* procedure. By using ISE, the processor modelling style may range from behavioral to RT-level models and may even partially extend to the gate level. The extracted register-transfer patterns do not depend on the detailed hardware structure. Efficiency of ISE is guaranteed by use of *Binary Decision Diagrams*.

With respect to customizable, application-specific DSPs, which are likely to replace off-the-shelf components in single-chip systems, the *short turnaround times* achieved in our approach are important. Retargeting the compiler to a new processor model, which includes ISE and tree parser generation, is possible within a few minutes of CPU time, even for complex target processors. In this way, a large part of the design space may be explored by the processor designer, before the final architecture gets fixed.

7.1.2 Code generation and optimization

Mapping of the intermediate program representation to processor-specific RT patterns relies on a standard technique from compiler construction, namely *covering of expression trees*. In this context, the contribution of this book is to show, that the processor models required by standard tree parser generators can be *automatically derived from HDL models*, which are common in hardware design.

Furthermore, we have focussed on the problem of *efficient memory address generation* for DSPs. The presented offset assignment algorithms for arranging program values in memory outperform those ones presented in earlier work. These algorithms are both easy to implement and runtime efficient. Also modify registers are systematically included into address optimization. Based on the generic AGU model defined for offset assignment problems, we have also proposed techniques for array addressing and exploitation of dedicated hardware for delay lines.

Finally a thorough investigation of the *code compaction* problem for DSPs was given. Related projects in embedded code generation so far rely on heuristic compaction algorithms developed in the area of microprogramming. Application studies showed that exploitation of potential parallelism at the instruction-level still is a difficult problem for DSPs. The *Integer Programming based compaction technique* presented in this book is capable of optimally exploiting local parallelism. In combination with a commercial IP solver this technique applies to problems of relevant size. The presented definition of the code compaction problem is more general than in previous work. As a result, its applicability could be demonstrated for a complex standard DSP, for which effective code compaction techniques have not been reported so far.

Conclusions 193

7.2 FUTURE RESEARCH

A number of interesting topics could not be treated here, but remain open for future work. The experiences gained with MSSQ and RECORD permit to identify promising areas of potential improvements, which could be achieved on the basis of the presented techniques.

Design of embedded systems based on programmable processors also demands for tool support beyond machine code generation. With increasing use of processor cores, also verification and **validation of core models** are of outstanding importance. This includes equivalence checking between the different synthesis and simulation models of a single core. During discussion of instruction-set extraction we have already outlined possible applications of ISE to model checking. Equivalence of instruction sets extracted from processor models differing in the level of abstraction, for instance, gives evidence of correct transformation of an instruction-set model into an RTL model and vice versa. Thus, ISE techniques may be reused for validation of core models. This would permit a higher integration of the different areas of model validation and code generation.

The issue of coupling code generation phases was mentioned several times throughout this book. A certain degree of **phase coupling** was achieved, which turned out to be essential for acceptable code quality. Remarkable improvements in code quality will demand for an even closer coupling of phases. The main challenge, however, is to cleverly restrict the resulting huge search space. A promising approach might be the use of "constraint logic programming" (CLP) [BiMa95], which automates the required backtracking mechanism. However, CLP has so far not been applied to generation of *optimized* code.

According to the experiences of commercial CAD tool vendors, pure "pushbutton" tools are hardly desired by the customers. Instead, the human designer wants to have visual control over produced results as well as some **opportunities for interaction**. Therefore, code generators intended to be integrated into VLSI design systems will benefit from comfortable graphical user interfaces. Such interfaces, for instance, could support manual binding of variables to registers or binding of operations to functional units. This leads to a concept of embedded code generation by means of an interactive development tool, which is in sharp contrast to the classical role of compilers, but would probably increase the acceptance of high-level programming languages in DSP system design.

REFERENCES

[ADK+95] G. Araujo, S. Devadas, K. Keutzer, S. Liao, S. Malik, A. Sudarsanam, S. Tjiang, A. Wang: *Challenges in Code Generation for Embedded Processors*, chapter 3 in [MaGo95]

[AGT89] A.V. Aho, M. Ganapathi, S.W.K Tjiang: *Code Generation Using Tree Matching and Dynamic Programming*, ACM Trans. on Programming Languages and Systems 11, no. 4, 1989, pp. 491-516

[AJU77] A.V. Aho, S.C. Johnson, J.D. Ullman: *Code Generation for Expressions with Common Subexpressions*, Journal of the ACM, vol. 24, no. 1, 1977

[AML96] G. Araujo, S. Malik, M. Lee: *Using Register Transfer Paths in Code Generation for Heterogeneous Memory-Register Architectures*, 33rd Design Automation Conference (DAC), 1996

[ASM96] G. Araujo, A. Sudarsanam, S. Malik: *Instruction Set Design and Optimizations for Address Computation in DSP Architectures*, 9th Int. Symp. on System Synthesis (ISSS), 1996

[ASU86] A.V. Aho, R. Sethi, J.D. Ullman: *Compilers - Principles, Techniques, and Tools*, Addison-Wesley, 1986

[AhJo76] A.V. Aho, S.C. Johnson: *Optimal Code Generation for Expression Trees*, Journal of the ACM, vol. 23, no. 3, 1976, pp. 488-501

[AhUl72] A.V. Aho, J.D. Ullman: *The Theory of Parsing, Translation and Compiling*, vols. I and II, Prentice Hall, 1972

[AiNi88] A. Aiken, A. Nicolau: *A Development Environment for Horizontal Microcode*, IEEE Trans. on Software Engineering, no. 14, 1988, pp. 584-594

[Alla90] V.H. Allan: *Peephole Optimization as a Targeting and Coupling Tool*, 22nd Annual Workshop on Microprogramming and Microarchitecture (MICRO-23), 1990, pp. 112-121

[Ana91] Analog Devices Inc.: *ADSP-2101/2102 User's Manual*, 1991

[ArMa95] G. Araujo, S. Malik: *Optimal Code Generation for Embedded Memory Non-Homogeneous Register Architectures*, 8th Int. Symp. on System Synthesis (ISSS), 1995, pp. 36-41

[BBH+94] S. Bashford, U. Bieker, B. Harking, R. Leupers, P. Marwedel, A. Neumann, D. Voggenauer: *The MIMOLA Language V4.1*, Technical Report, University of Dortmund, Dept. of Computer Science, September 1994

[BBM86] M. Balakrishnan, P.C.P. Bhatt, B.B. Madan: *An Efficient Retargetable Microcode Generator*, 19th Ann. Workshop on Microprogramming (MICRO-19), 1986, pp. 44-53

[BCT92] P. Briggs, K.D. Cooper, L. Torczon: *Rematerialization*, ACM SIGPLAN Conference on Programming Language Design and Implementation (PLDI), 1992, pp. 311-321

[BDB90] A. Balachandran, D.M. Dhamdere, S. Biswas: *Efficient Retargetable Code Generation Using Bottom-Up Tree Pattern Matching*, Comput. Lang. vol. 15, no. 3, 1990, pp. 127-140

[BEH91] D. Bradlee, S. Eggers, R. Henry: *Integrating Register Allocation and Instruction Scheduling for RISCs*, 4th Int. Conf. on Architectural Support for Programming Languages and Operating Systems (ASPLOS), 1991

[BHE91] D.G. Bradlee, R.R. Henry, S.J. Eggers: *The Marion System for Retargetable Instruction Scheduling*, ACM SIGPLAN Conference on Programming Language Design and Implementation (PLDI), 1991, pp. 229-235

[BRB90] K.S. Brace, R.L. Rudell, R.E. Bryant: *Efficient Implementation of a BDD Package*, 27th Design Automation Conference (DAC), 1990, pp. 40-45

[BSBC95] T.S. Brasier, P.H. Sweany, S. Carr, S.J. Beaty: *CRAIG: A Practical Framework for Combining Instruction Scheduling and Register Allocation*, Int. Conf. on Parallel Architectures and Compilation Techniques (PACT), 1995, pp. 11-18

[BSV93] K. Buchenrieder, A. Sedlmeier, C. Veith: *Design of HW/SW Systems with VLSI Subsystems Using CODES*, 6th IEEE Workshop on VLSI Signal Processing, 1993, pp. 233-239

[BaHa81] T. Baba, H. Hagiwara: *The MPG System: A Machine Independent Efficient Microprogram Generator*, IEEE Trans. on Computers, vol. 30, no. 6, 1981, pp. 373-395

[Bane93] U. Banerjee: *Loop Transformations for Restructuring Compilers – The Foundations*, Kluwer Academic Publishers, 1993

[Bart92] D.H. Bartley: *Optimizing Stack Frame Accesses for Processors with Restricted Addressing Modes*, Software – Practice and Experience, vol. 22(2), 1992, pp. 101-110

[Bash95] S. Bashford: *Code Generation Techniques for Irregular Architectures*, Technical Report no. 596, Dept. of Computer Science, University of Dortmund, Germany, 1995

REFERENCES

[Bela66] L.A. Belady: *A Study of Replacement Algorithms for a Virtual-Storage Computer*, IBM System Journals 5(2): pp. 78-101, 1966

[BiMa95] U. Bieker, P. Marwedel: *Retargetable Self-Test Program Generation Using Constraint Logic Programming*, 32nd Design Automation Conference (DAC), 1995, pp. 605-611

[Biek95] U. Bieker: *Retargierbare Compilierung von Selbsttestprogrammen digitaler Prozessoren mittels Constraint-logischer Programmierung* (in German), Doctoral thesis, Dept. of Computer Science, University of Dortmund, Germany, Shaker Verlag, 1995

[BoGi77] F.T. Boesch, J.F. Gimpel: *Covering the Points of a Digraph with Point-Disjoint Paths and Its Application to Code Optimization*, Journal of the ACM, vol. 24, no. 2, 1977, pp. 192-198

[BrSe76] J. Bruno, R. Sethi: *Code Generation for a One-Register Machine*, Journal of the ACM, no. 23, 1976, pp. 502-510

[Brig92] P. Briggs: *Register Allocation via Graph Coloring*, Doctoral thesis, Dept. of Computer Science, Rice University, Houston/Texas, 1992

[Brya85] R.E. Bryant: *Symbolic Manipulation of Boolean Functions Using a Graphical Representation*, 22nd Design Automation Conference (DAC), 1985, pp. 688-694

[Brya86] R.E. Bryant: *Graph-based Algorithms for Boolean Function Manipulation*, IEEE Trans. on Computers, 40, no. 2, 1986, pp. 205-213

[Brya92] R.E. Bryant: *Symbolic Boolean Manipulation with Ordered Binary Decision Diagrams*, ACM Computing Surveys, vol. 24, no. 3, 1992, pp. 293-318

[Calv93] J.P. Calvez: *Embedded Real-Time Systems*, John Wiley & Sons, 1993

[Catt78] R.G.G. Cattell: *Formalization and Automatic Derivation of Code Generators*, Doctoral thesis, Dept. of Computer Science, Carnegie-Mellon University, Pittsburgh, 1978

[ChBo94] P. Chou, G. Borriello: *Software Scheduling in the Co-Synthesis of Reactive Real-Time Systems*, 31st Design Automation Conference (DAC), 1994, pp. 1-4

[Chai82] G.J. Chaitin: *Register Allocation and Spilling via Graph Coloring*, SIGPLAN Symp. on Compiler Construction, 1982, pp. 98-105

[DFL93] Mentor Graphics Corporation: *DSP Architect DFL User's and Reference Manual, V 8.2_6*, 1993

[DLH88] D.J. DeFatta, J.G. Lucas, W.S. Hodgekiss: *Digital Signal Processing: A System Design Approach*, John Wiley & Sons, 1988

[DLSM81] S. Davidson, D. Landskov, B.D. Shriver, P.W. Mallett: *Some Experiments in Local Microcode Compaction for Horizontal Machines*, IEEE Trans. on Computers, vol. 30, no. 7, 1981, pp. 460-477

[DaFr84] J.W. Davidson, C.W. Fraser: *Automatic Generation of Peephole Optimizers*, ACM SIGPLAN Notices, vol. 19, no. 6, 1984, pp. 111-116

[DeWi76] D.J. DeWitt: *A Machine Independent Approach to the Production of Optimal Horizontal Microcode*, Doctoral thesis, Technical Report 76 DT 4, University of Michigan, 1976

[EHB93] R. Ernst, J. Henkel, T. Benner: *Hardware-Software Cosynthesis for Microcontrollers*, IEEE Design & Test Magazine, no. 12, 1993, pp. 64-75

[ESL89] H. Emmelmann, F.W. Schröer, R. Landwehr: *BEG – A Generator for Efficient Backends*, ACM SIGPLAN Conference on Programming Language Design and Implementation (PLDI), SIGPLAN Notices 24, no. 7, 1989, pp. 227-237

[Emme93] H. Emmelmann: *Codeselektion mit regulär gesteuerter Termersetzung* (in German), Doctoral thesis, Dept. of Computer Science, University of Karlsruhe, Germany, 1993

[FHP92a] C.W. Fraser, D.R. Hanson, T.A. Proebsting: *Engineering a Simple, Efficient Code Generator Generator*, ACM Letters on Programming Languages and Systems, vol. 1, no. 3, 1992, pp. 213-226

[FHP92b] C.W. Fraser, R.R. Henry, T.A. Proebsting: *BURG – Fast Optimal Instruction Selection and Tree Parsing*, ACM SIGPLAN Notices 27 (4), 1992, pp. 68-76

[FaKn93] A. Fauth, A. Knoll: *Translating Signal Flowcharts into Microcode for Custom Digital Signal Processors*, Int. Conf. on Signal Processing (ICSP), 1993, pp. 65-68

[FHKM94] A. Fauth, G. Hommel, A. Knoll, C. Müller: *Global Code Selection for Directed Acyclic Graphs*, in: P.A. Fritzson (ed.): 5th Int. Conference on Compiler Construction, 1994

[FVM95] A. Fauth, J. Van Praet, M. Freericks: *Describing Instruction-Set Processors in nML*, European Design and Test Conference (ED & TC), 1995, pp. 503-507

[Fish81] J.A. Fisher: *Trace Scheduling: A Technique for Global Microcode Compaction*, IEEE Trans. on Computers, vol. 30, no. 7, 1981, pp. 478-490

[GCLD92] G. Goossens, F. Catthoor, D. Lanneer, H. De Man: *Integration of Signal Processing Systems on Heterogeneous IC Architectures*, 5th High-Level Synthesis Workshop (HLSW), 1992, pp. 16-26

[GLV96] W. Geurts, D. Lanneer, J. Van Praet, et al.: *Design of DSP Systems with CHESS/CHECKERS*, Handouts of the 2nd Int. Workshop on Embedded Code Generation, Leuven/Belgium, March 1996

REFERENCES

[GVNG94] D. Gajski, F. Vahid, S. Narayan, J. Gong: *Specification and Design of Embedded Systems*, Prentice Hall, 1994

[GFH82] M. Ganapathi, C.N. Fischer, J.L. Hennessy: *Retargetable Compiler Code Generation*, ACM Computing Surveys, vol. 14, 1982, pp. 573-592

[GaJo79] M.R. Gary, D.S. Johnson: *Computers and Intractability – A Guide to the Theory of NP-Completeness*, Freemann, 1979

[GaVa95] D. Gajski, F. Vahid: *Specification and Design of Embedded Hardware-Software Systems*, IEEE Design & Test of Computers, Spring 1995, pp. 53-67

[GeEl92] C. Gebotys, M. Elmasry: *Optimal VLSI Architectural Synthesis*, Kluwer Academic Publishers, 1992

[Glan77] R.S. Glanville: *A Machine Independent Algorithm for Code Generation and its Use in Retargetable Compilers*, Doctoral thesis, University of California at Berkeley, 1977

[GoHs88] J. Goodman, W. Hsu: *Code Scheduling and Register Allocation in Large Basic Blocks*, ACM SIGPLAN Conference on Programming Language Design and Implementation (PLDI), 1988

[Golu80] M.C. Golumbic: *Algorithmic Graph Theory and Perfect Graphs*, Academic Press, 1980

[Goos96] G. Goossens: *Code Generation for Embedded Processors – Introductory Talk*, Handouts of the 2nd Int. Workshop on Embedded Code Generation, Leuven/Belgium, March 1996

[GuDe92] R.K. Gupta, G. De Micheli: *System-Level Synthesis Using Re-Programmable Components*, European Conference on Design Automation (EDAC), 1992, pp. 2-8

[Gupt95] R.K. Gupta: *Co-Synthesis of Hardware and Software for Digital Embedded Systems*, Kluwer Academic Publishers, 1995

[Hart92] R. Hartmann: *Combined Scheduling and Data Routing for Programmable ASIC Systems*, European Conference on Design Automation (EDAC), 1992, pp. 486-490

[HeGl94] M. Held, M. Glesner: *Generating Compilers for Generated Datapaths*, European Design Automation Conference (EURO-DAC), 1994, pp. 532-537

[HePa90] J.L. Hennessy, D.A. Patterson: *Computer Architecture – A Quantitative Approach*, Morgan Kaufmann Publishers Inc., 1990

[Hilf85] P. Hilfinger: *A High-Level Language and Silicon Compiler for Digital Signal Processing*, Custom Integrated Circuits Conference (CICC), 1985, pp. 213-316

[HoSa87] E. Horowitz, S. Sahni: *Fundamentals of Data Structures in PASCAL*, 2nd Edition, Computer Science Press Inc., 1987

[Hwan93] K. Hwang: *Advanced Computer Architecture*, McGraw-Hill, 1993

[IEEE88] IEEE Design Automation Standards Subcommittee: *IEEE Standard VHDL Language Reference Manual*, IEEE Std. 1076-1987, IEEE Inc., New York, 1988

[Inte96] Intermetrics: *NEC 77016 DSP C Compiler – Product Description*, Intermetrics Microsystems Software Inc., Cambridge (Mass.), 1996

[JoAl90] R.B. Jones, V.H. Allan: *Software Pipelining: A Comparison and Improvement*, 22nd Annual Workshop on Microprogramming and Microarchitecture (MICRO-23), 1990, pp. 46-56

[KAJW96] S. Kumar, J.H. Aylor, B.W. Johnson, W.A. Wulf: *The Codesign of Embedded Systems*, Kluwer Academic Publishers, 1996

[KaLe93] A. Kalavade, E.A. Lee: *A Hardware-Software Codesign Methodology for DSP Applications*, IEEE Design & Test Magazine, no. 9, 1993, pp. 16-28

[Krug91] G. Krüger: *A Tool for Hierarchical Test Generation*, IEEE Trans. on CAD, vol. 10, no. 4, 1991, pp. 519-524

[KuPa87] F.J. Kurdahi, A.C. Parker: *REAL: A Program for Register Allocation*, 24th Design Automation Conference (DAC), 1987, pp. 210-215

[LCGD94] D. Lanneer, M. Cornero, G. Goossens, H. De Man: *Data Routing: A Paradigm for Efficient Data-Path Synthesis and Code Generation*, 7th Int. Symp. on High-Level Synthesis (HLSS), 1994, pp. 17-21

[LDK+95a] S. Liao, S. Devadas, K. Keutzer, S. Tjiang, A. Wang: *Storage Assignment to Decrease Code Size*, ACM SIGPLAN Conference on Programming Language Design and Implementation (PLDI), 1995

[LDK+95b] S. Liao, S. Devadas, K. Keutzer, S. Tjiang, A. Wang: *Code Optimization Techniques for Embedded DSP Microprocessors*, 32nd Design Automation Conference (DAC), 1995, pp. 599-604

[LDK+95c] S. Liao, S. Devadas, K. Keutzer, S. Tjiang: *Instruction Selection Using Binate Covering for Code Size Optimization*, Int. Conf. on Computer-Aided Design (ICCAD), 1995, pp. 393-399

[LMD94] B. Landwehr, P. Marwedel, R. Dömer: *OSCAR: Optimum Simultaneous Scheduling, Allocation, and Resource Binding based on Integer Programming*, European Design Automation Conference (EURO-DAC), 1994

[LMP94a] C. Liem, T. May, P. Paulin: *Instruction-Set Matching and Selection for DSP and ASIP Code Generation*, European Design and Test Conference (ED & TC), 1994, pp. 31-37

REFERENCES

[LMP94b] C. Liem, T. May, P. Paulin: *Register Assignment through Resource Classification for ASIP Microcode Generation*, Int. Conf. on Computer-Aided Design (ICCAD), 1994, pp. 397-402

[LNM94] R. Leupers, R. Niemann, P. Marwedel: *Methods for Retargetable DSP Code Generation*, 7th IEEE Workshop on VLSI Signal Processing, 1994, pp. 127-136

[LPCJ95] C. Liem, P. Paulin, M. Cornero, A. Jerraya: *Industrial Experience Using Rule-driven Retargetable Code Generation for Multimedia Applications*, 8th Int. Symp. on System Synthesis (ISSS), 1995, pp. 60-65

[LPJ96] C. Liem, P. Paulin, A. Jerraya: *Address Calculation for Retargetable Compilation and Exploration of Instruction-Set Architectures*, 33rd Design Automation Conference (DAC), 1996

[LPJ97] C. Liem, P. Paulin, A. Jerraya: *ReCode: The Design and Redesign of the Instruction Codes for Embedded Instruction-Set Processors*, European Design and Test Conference (ED & TC), 1997, p. 612

[LSM94] R. Leupers, W. Schenk, P. Marwedel: *Retargetable Assembly Code Generation by Bootstrapping*, 7th Int. Symp. on High-Level Synthesis (HLSS), 1994, pp. 88-93

[LSU89] R. Lipsett, C. Schaefer, C. Ussery: *VHDL: Hardware Description and Design*, Kluwer Academic Publishers, 1989

[LVKS+95] D. Lanneer, J. Van Praet, A. Kifli, K. Schoofs, W. Geurts, F. Thoen, G. Goossens: *CHESS: Retargetable Code Generation for Embedded DSP Processors*, chapter 5 in [MaGo95]

[LaCe93] M. Langevin, E. Cerny: *An Automata-Theoretic Approach to Local Microcode Generation*, European Conference on Design Automation (EDAC), 1993, pp. 94-98

[Lam88] M. Lam: *Software Pipelining: An Effective Scheduling Technique for VLIW machines*, ACM SIGPLAN Conference on Programming Language Design and Implementation (PLDI), 1988, pp. 318-328

[LeMa94] R. Leupers, P. Marwedel: *Instruction Set Extraction from Programmable Structures*, European Design Automation Conference (EURO-DAC), 1994, pp. 156-161

[LeMa97] R. Leupers, P. Marwedel: *Retargetable Code Generation based on Structural Processor Descriptions*, Journal on Design Automation for Embedded Systems, Kluwer Academic Publishers, 1997

[LeMe88] E.A. Lee, D.G. Messerschmitt: *Digital Communication*, Kluwer Academic Publishers, 1988

[Lee88] E.A. Lee: *Programmable DSP Architectures*, Part I: IEEE ASSP Magazine, October 1988, pp. 4-19, Part II: IEEE ASSP Magazine, January 1989, pp. 4-14

[Liao96] S. Liao: *Code Generation and Optimization for Embedded Digital Signal Processors*, Doctoral thesis, Dept. of Electrical Engineering and Computer Science, Massachusetts Institute of Technology, 1996

[MME90] M. Mahmood, F. Mavaddat, M.I. Elmasry: *Experiments with an Efficient Heuristic Algorithm for Local Microcode Generation*, Int. Conf. on Computer Design (ICCD), 1990, pp. 319-323

[MaEw94] C. Marven, G. Ewers: *A simple Approach to Digital Signal Processing*, Texas Instruments, 1994

[MaGo95] P. Marwedel, G. Goossens (eds.): *Code Generation for Embedded Processors*, Kluwer Academic Publishers, 1995

[MaSc93] P. Marwedel, W. Schenk: *Cooperation of Synthesis, Retargetable Code Generation and Test Generation in the MIMOLA Software System*, European Conference on Design Automation (EDAC), 1993, pp. 63-69

[MaTe96] F. Mavaddat, A. Teimoortagh: *Using Recursive Descent Parsing to Generate Retargetable Microcode with Memory References*, Handouts of the 2nd Int. Workshop on Embedded Code Generation, Leuven/Belgium, March 1996

[Mahm96] M. Mahmood: *Formal Language Approach to Retargetable Microcode Synthesis*, Handouts of the 2nd Int. Workshop on Embedded Code Generation, Leuven/Belgium, March 1996

[Mano93] M.M. Mano: *Computer System Architecture*, 3rd Edition, Prentice Hall, 1993

[Marw93] P. Marwedel: *Tree-based Mapping of Algorithms to Predefined Structures*, Int. Conf. on Computer-Aided Design (ICCAD), 1993, pp. 586-993

[Marw95] P. Marwedel: *Code Generation for Embedded Processors: An Introduction*, chapter 1 in [MaGo95]

[MoBr95] C. Monahan, F. Brewer: *Symbolic Modelling and Evaluation of Data Paths*, 32nd Design Automation Conference (DAC), 1995, pp. 389-394

[Moto92] Motorola Inc.: *DSP 56156 Digital Signal Processor User's Manual*, 1992

[MuVa83] R.A. Mueller, J. Varghese: *Flow Graph Machine Models in Microcode Synthesis*, 16th Ann. Workshop on Microprogramming (MICRO-16), 1983, pp. 159-167

[NND95] S. Novack, A. Nicolau, N. Dutt: *A Unified Code Generation Approach using Mutation Scheduling*, chapter 12 in [MaGo95]

REFERENCES

[NiMa96] R. Niemann, P. Marwedel: *Hardware/Software Partitioning Using Integer Programming*, European Design and Test Conference (ED & TC), 1996, pp. 473-479

[NiPo91] A. Nicolau, R. Potasman: *Incremental Tree Height Reduction for High-Level Synthesis*, 28th Design Automation Conference (DAC), 1991, pp. 770-774

[Nowa87a] L. Nowak: *Graph based Retargetable Microcode Compilation in the MIMOLA Design System*, 20th Ann. Workshop on Microprogramming (MICRO-20), 1987, pp. 126-132

[NoMa89] L. Nowak, P. Marwedel: *Verification of Hardware Descriptions by Retargetable Code Generation*, 26th Design Automation Conference (DAC), 1989, pp. 441-447

[OpSc75] A.V. Oppenheim, R.W. Schafer: *Digital Signal Processing*, Prentice Hall, 1975, 2nd edition 1988

[PCL+96] P. Paulin, M. Cornero, C. Liem, et al.: *Trends in Embedded Systems Technology*, in: M.G. Sami, G. De Micheli (eds.): *Hardware/Software Codesign*, Kluwer Academic Publishers, 1996

[PLMS92] P. Paulin, C. Liem, T. May, S. Sutarwala: *DSP Design Tool Requirements for the Nineties: An Industrial Perspective*, Technical Report, Bell Northern Research, 1992

[PLMS95] P. Paulin, C. Liem, T. May, S. Sutarwala: *FlexWare: A Flexible Firmware Development Environment for Embedded Systems*, in [MaGo95]

[RaGo75] L.R. Rabiner, B. Gold: *Theory and Application of Digital Signal Processing*, Prentice Hall, 1975

[RiHi88] K. Rimey, P.N. Hilfinger: *Lazy Data Routing and Greedy Scheduling for Application-Specific Signal Processors*, 21st Annual Workshop on Microprogramming and Microarchitecture (MICRO-21), 1988, pp. 111-115

[RoSe90] N. Robertson, P.D. Seymour: *An Outline of Disjoint Path Algorithms*, pp. 267-292 in: B. Korte, L. Lovasz, H.J. Prömel, A. Schrijver (eds.): *Paths, Flows, and VLSI Layout*, Springer-Verlag, 1990

[SMT+95] M. Strik, J. van Meerbergen, A. Timmer, J. Jess, S. Note: *Efficient Code Generation for In-House DSP Cores*, European Design and Test Conference (ED & TC), 1995, pp. 244-249

[SWB96] J. Shu, T. Wilson, D. Banerji: *Instruction-Set Matching and GA-based Selection for Embedded-Processor Code Generation*, 9th Int. Conf. on VLSI Design, 1996, pp. 73-76

[Sche95] W. Schenk: *Retargetable Code Generation for Parallel, Pipelined Processor Structures*, chapter 7 in [MaGo95]

[Seth75] R. Sethi: *Complete Register Allocation Problems*, SIAM J. Computing 4(3), 1975, pp. 226-248

[Stal93] R.M. Stallmann: *Using and Porting GNU CC* V2.4, Free Software Foundation, Cambridge/Massachusetts, 1993

[SuMa95] A. Sudarsanam, S. Malik: *Memory Bank and Register Allocation in Software Synthesis for ASIPs*, Int. Conf. on Computer-Aided Design (ICCAD), 1995, pp. 388-392

[TI90] Texas Instruments: *TMS320C2x User's Guide*, rev. B, 1990

[TSMJ95] A. Timmer, M. Strik, J. van Meerbergen, J. Jess: *Conflict Modelling and Instruction Scheduling in Code Generation for In-House DSP Cores*, 32nd Design Automation Conference (DAC), 1995, pp. 593-598

[Tane90] A.S. Tanenbaum: *Structured Computer Organization*, 3rd Edition, Prentice Hall, 1990

[VGLD94] J. Van Praet, G. Goossens, D. Lanneer, H. De Man: *Instruction Set Definition and Instruction Selection for ASIPs*, 7th Int. Symp. on High-Level Synthesis (HLSS), 1994, pp. 11-16

[VLG+96] J. Van Praet, D. Lanneer, G. Goossens, W. Geurts, H. De Man: *A Graph Based Processor Model for Retargetable Code Generation*, European Design and Test Conference (ED & TC), 1996

[VVE+95] P. Vanoostende, E. Vanzieleghem, E. Rousseau, C. Massy, F. Gerard: *Retargetable Code Generation: Key Issues for Successful Introduction*, chapter 2 in [MaGo95]

[Vegd82a] S.R. Vegdahl: *Local Code Generation and Compaction in Optimizing Microcode Compilers*, Doctoral thesis, Dept. of Computer Science, Carnegie-Mellon University, 1982

[Vegd82b] S.R. Vegdahl: *Phase Coupling and Constant Generation in an Optimizing Microcode Compiler*, 15th Ann. Workshop on Microprogramming (MICRO-15), 1982, pp. 125-133

[WGHB94] T. Wilson, G. Grewal, B. Halley, D. Banerji: *An Integrated Approach to Retargetable Code Generation*, 7th Int. Symp. on High-Level Synthesis (HLSS), 1994, pp. 70-75

[Wess92] B. Wess: *Automatic Instruction Code Generation based on Trellis Diagrams*, IEEE Int. Symp. on Circuits and Systems (ISCAS), 1992, pp. 645-648

[Wess96] B. Wess: *Translating Expression DAGs into Optimized Code for non-homogeneous Register Machines*, Handouts of the 2nd Int. Workshop on Embedded Code Generation, Leuven/Belgium, March 1996

REFERENCES

[WiMa95] R. Wilhelm, D. Maurer: *Compiler Design*, Addison-Wesley, 1995

[ZTM95] V. Zivojnovic, S. Tjiang, H. Meyr: *Compiled Simulation of Programmable DSP Architectures*, IEEE Workshop on VLSI Signal Processing, 1995, pp. 187-196

[ZVSM94] V. Zivojnovic, J.M. Velarde, C. Schläger, H. Meyr: *DSPStone – A DSP-oriented Benchmarking Methodology*, Int. Conf. on Signal Processing Applications and Technology (ICSPAT), 1994, also available as Technical Report, Dept. of Electrical Engineering, Institute for Integrated Systems for Signal Processing, University of Aachen, Germany

INDEX

Access graph, 134, 146
Access sequence, 133, 139
Address assignment, 92, 127, 131, 133, 135, 173
Address generation unit, 10, 12, 128, 131, 156
Address register assignment, 152
Address register, 130
AGU operations, 131
ALAP, 163
Array address mapping, 148
Array index expression, 147
ASAP, 163
ASIP, 13, 36, 81, 180
Auto-increment/decrement, 128, 133
Auto-modify, 130, 142
Background register, 93, 100, 128
Basic block, 24, 87, 101, 162
Behavioral analysis, 60
Behavioral model, 45
Behavioral module, 29
Binary decision diagram, 51, 57, 79
Bit vector, 30
Bit-true, 16, 96
Boolean function, 50
Bootstrapping, 41
Bus, 165
Chaining, 81, 113
Code compaction, 24, 36, 92, 162, 164, 168
Code generation phases, 91
Code generator generator, 23, 88, 107

Code quality, 8, 11, 14, 27, 40, 42, 113, 175, 184, 189
Code selection, 19, 21, 86, 91
Combinational module, 30
Common subexpression, 87
Compaction strategies, 177
Compilation speed, 15, 40, 108, 113, 173, 189
Conflict representation, 164
Connection operation graph, 33–34
Connections, 31
Contiguous schedule, 120
Control port, 48
Control signal, 48, 79
Control step, 20, 162
Control-dependence, 100
Control/data-flow graph, 19
Controller, 17
Converse operators, 116
Core, 3, 12
Critical path, 163–164
Data routing, 25, 90
Data-anti-dependence, 100, 119, 170
Data-dependence, 100, 119, 170
Data-flow description, 16
Data-flow graph, 87
Data-flow tree, 87
Decision variables, 169
Delay line, 9, 11, 16, 39, 96, 155
Derivation, 106
DFL data types, 96
DFL language, 15, 95
Digital filtering, 8, 36, 96
Digital signal processing, 8

Direct addressing, 130
DSP classification, 13
DSP processor architecture, 10
Dynamic condition variable, 55, 62
Dynamic condition, 54
Dynamic programming, 107
Embedded code generation, 8, 14, 25
Embedded system, 2
Evaluation order, 119
Expression tree assignment, 99, 101
Expression tree, 99
FIR filter, 9, 97
First-come-first-served, 164
Foreground register, 93, 121, 127
General offset assignment, 139–140, 174
General-purpose system, 2
GNU C compiler, 23
Guarded expression, 67
Guarded port assignment, 62
Guarded register transfer pattern, 55
Guarded tristate assignment, 62
Guarded variable assignment, 62
Hardware description language, 17
Hardware multiplier, 10
Hardwired constant, 71
Heterogeneous system, 3
Hidden instruction, 83
Homogeneous architecture, 85
Horizontal side effect, 166
HW/SW codesign, 3
HW/SW partitioning, 3
I-tree, 34
Iburg, 23, 107
Indirect addressing, 130
Induction variable, 146
Inhomogeneous architecture, 86
Instruction bit variable, 54
Instruction decoder, 37

Instruction format, 11, 14, 23, 37, 164
Instruction-level parallelism, 11, 21, 24, 86, 162
Instruction-set extraction, 47, 55, 80
Instruction-set model, 52
Integer Programming, 26, 91, 168
Integrated code generation, 90
Inter-iteration distance graph, 150
Inter-RT conflict, 171
Interference graph, 86
Intermediate program representation, 19, 86, 93, 98
Intra-iteration distance graph, 149
Intra-RT conflict, 58
Jump address insertion, 173
Length register, 156
Lexical analysis, 18
Lifetime, 85, 170
List scheduling, 164
Load-store architecture, 90
Local condition, 62, 75
Longest path heuristic, 152
Loop folding, 25, 187
Loop unrolling, 25
Loop variable, 146
Loop-invariant code motion, 146
Loops, 104, 146
LR parsing, 22
Microinstruction, 24
Microoperation, 17, 23
Microprogramming, 17, 23
MIMOLA, 17, 31, 33, 38
Mixed-level description, 37, 46
Mobility range, 169
Mode register variable, 55, 61
Mode register, 10–11, 23, 54–56, 123, 165
Model validation, 82, 193
Modify register, 130, 141
Module port, 30

Module variable, 30
Modulo addressing, 156
Modulus logic, 157
MSSQ, 25, 33, 43, 89
Multiply-accumulate, 9, 11, 21, 166, 187
Mutual exclusion, 82
Nested loop, 154
NOP, 36, 58, 78, 169–170, 172
NOP-variables, 169
NP-hardness, 20
Off-the-shelf DSP, 1, 13
Operator mismatch, 115
Output-dependence, 100, 119, 170
Overall distance graph, 150
Parallel schedule, 162
Partial instruction, 34, 164
Path covering, 151
Pattern matching, 34, 105
Peephole optimization, 23
Phase coupling, 20, 23
Phase ordering, 20
Port variable, 61
Post-modify, 128
Processor description style, 46
Processor graph model, 59, 70
Program value, 99
Read set, 119
Real-time constraint, 166, 176, 181
RECORD compiler, 179
Recursive descent, 89
Redundant variables, 171
Register allocation, 19, 21, 85, 92
Register binding, 92, 100
Register deadlock, 119
Register transfer condition, 54, 75, 79
Register transfer expression, 53, 72
Register transfer level, 31
Register transfer pattern, 52
Register transfer, 17, 23
Reservation, 33

Retargetability, 5, 14, 27, 179
Ring buffer, 11, 155
Rounding, 96
RT dependency graph, 162
RT pattern selection, 92
RT scheduling, 92, 118, 121
RT-level basic block, 87
Rule-based code generation, 90
Sample period, 9
Saturating arithmetic, 96
Scheduling, 20, 86, 92, 94, 118
Sequential module, 30
Sequential schedule, 119
Shared control signal, 57
Side effect, 36, 165, 170, 173
Signal flow graph, 9, 16, 97
SILAGE language, 15
Simple loop, 147
Simple offset assignment, 134, 137
Single assignment form, 86–87
Single-chip system, 3
Source-level basic block, 87
Source-level scheduling, 92
Special-purpose register, 10–11, 86
Spilling, 20, 94, 121
Standard optimizations, 19
Strength reduction, 19, 146
Structural analysis, 68
Structural mismatch, 114
Structural model, 41, 46, 59
Structural module, 29
Support set, 75
Syntax analysis, 18
System-level design automation, 1
Test program generator, 23
Texas Instruments DSP, 11, 41, 81, 125, 128, 173
Tie-break function, 136
Transformation rules, 88, 94, 113, 118
Transition frequency, 134
Tree cover, 105

Tree covering, 88
Tree grammar specification, 107
Tree grammar, 106
Tree language, 106
Tree parser generation, 107, 110, 112
Tree parser, 107
Tree parsing, 105
Tree selection, 92
Tristate, 36, 56, 75
Undefined condition, 77, 79
Undefined output, 83
Update value, 148
Version selection, 92, 165, 169
Version shuffling, 165, 169–170, 172
Version variables, 169
Version, 34, 165
Vertical code, 24, 92, 94, 124, 174
Vertical side effect, 166
VHDL, 17
Word-length effects, 11, 96
Write location, 119
Y-chart, 1
Zero-overhead loops, 104